DESIGN ENGINEERING –
sustainable and holistic

avedition

Marina-Elena Wachs

Preface

What do we need for 'sketching' our future?

'Firstly, the human, nature's servant and interpreter, uses and knows as much as he does about the order of nature through trials and observation; other than that, he knows nothing and is capable of nothing. Secondly, neither the bare hand, nor the mind – when left to itself – is capable of very much, […] And, in the same manner that tools steer and control the actions of the hand, the tools of the mind support and protect knowledge.' (Bacon, F., p. 26)[1]

This citation, from 1830, could just as well refer to the present time in Europe and the world, as the interconnection between the mind and the hand is equally relevant today.

At the beginning of the 2020s, we are discussing great visions for the future – the post-digital era – and the meaning of terms like 'society of knowledge'. But if we look at the philosopher Roland Barthes, who was cited in several books after his death, **he questioned the value and ownership of knowledge** by saying '*Interdisciplinarity consists in creating a new object that belongs to no one*'. But is that true?

His thesis and perhaps aim was to show the world the method for attaining new ways of perception, by addressing the impact of interdisciplinarity: On the one hand, innovation, to develop new ideas in relation to different realities, which could be the aim when creating a new object or innovative design concept. But, on the other hand, the reality today is that you know that 'you don't know anything', that relays to Francis Bacon (see above) and what is discussed today again.

Interdisciplinarity in the 21st century means using your 'base' and selecting multidisciplinary fields with which to collaborate, to get a holistic view for achieving results with optimised business and cultural benefits – in a sustainable manner.

Within the postmodern world and from a cultural studies' point of view, different cultural turns (see Bachmann-Medick, D., ibid: Nünning, A.) underline **the shift in the use of objects and materials, relationships, methods and languages in terms of 'codes' and 'parole'** (Ferdinand de Saussures). Today, we can observe these shifts in cultural and material behaviour, which mirror a more theory-based usage of these elements.

As such, we are no longer dealing with objects, but rather immateriality. Immateriality is the key factor in communicating content: using social media, communication and buying via the Internet of Things (IoT), thinking about design concepts and then producing a prototype by hand and with real 3D materials.

[1] Bacon, Franz, 1962 (original 1830), Neues Organ der Wissenschaften, Kapitel Aphorismen: Erstes Buch, Wissenschaftliche Buchgesellschaft, Darmstadt, p. 26, translated by Louise Huber-Fennell, 2021.

Thoughts and codes are our future capital. This development reflects the ongoing design engineering processes in industries, the methods used in designing and learning landscapes and programs at universities, as well as visionary enterprises: more and more design theoretical and systems-orientated design processes represent design. Like Donella Meadows, an environmental scientist, advised when 'testing' – or as I interpret as it, 'evaluating' – the value of a model, it is helpful to ask the following questions, in order to develop an innovative design concept.

Meadows's questions: 1. Are the driving factors likely to unfold this way?; 2. If they did, would the system react this way?; and 3. (the most important question) **What is driving the driving factors?** (Meadows, H. 2008, p. 47). Meadows introduced this model in her book 'Thinking in Systems'. It is important to remember: you never know, what will happen during the design process! It is only a basic model upon which to build design thinking models (interactive design models and others), to initiate design theoretical processes, and you never know which parameters will influence you along the way, or which coincidental key findings will be discovered during the process.

The behavioural change we need today – whether pertaining to sustainability or digitalisation with economic and social benefits – must follow this same path: What is driving the driving factor? However, the key parameters for change within the 21st century relates not only to the **drivers** – or trigger points – but also **motivation**. The philosopher Corine Pelluchon already pointed out the need for action (see Pelluchon, C., 2019) and **to question the motivation factors**. We have to act – not just speak about the shift or discuss sustainable solutions. This must be combined with the effective implementation of artificial intelligence for humans in the future, together with the use of methods that are attractive and acceptable to humans. The focal point is 'design doing', together with the transfer of knowledge in the post-digital era and creating a society of knowledge.

My imprint in the creative fields as a master tailor and as an industrial designer is founded on **interdisciplinary design activities** from the beginning of my studies, in the 1980s and 1990s, onwards: sketching with pencil and materials. In the process of writing this postdoctoral thesis, it has become clear to me there is a common understanding between designers of that generation: those who have the educational background, combined with handcraft apprenticeships and studied design at universities in Europe. Most of the universities in the Northern Countries were transfusing a common understanding. This was achieved through practise-based education, with a consciousness for design history and a vivid design discourse, which took place in theoretical courses and in interdisciplinary cooperation processes across diverse university subject fields, and included affiliated external stakeholders.

The design shift that we are recognizing today, is also influenced by the perspectives of various disciplines, ranging from engineering, material sciences and other scientific sectors, to philosophy. Philosophers like Roland Barthes and Michel Foucault provide important reading material that is relevant again right now!

And their insights explain this 'new' behaviour (and consequences) in this 'subject-object world', in relation to humans and machines. (see graphic 06: 'Interplay of different stakeholders (interests)', p. 169)

Today, when we speak about the power of Artificial Intelligence (AI) and cultural intelligence, we also have to consider emotional and social intelligence. And, if we look back at the dominant model of 'emotional intelligence' (EQ), conceived by the psychologist Daniel Goleman, and the model of 'Cultural Intelligence Individual Interactions Across Cultures' that was subsequently developed in the 1990s by Christopher Earley and Soon Ang, it currently finds renewed relevance. *'The major areas of skill a person with emotional intelligence possesses include knowing one's emotions, such as knowing a feeling as it occurs; managing emotions, such as the ability to soothe oneself, motivating oneself; […].'*[2] Imagine how much cultural education and the economy can profit from emotional memories formed at pre-school level. If we reread literature, such as those written by Roland Barthes (see chapter 2) or 'Die Ordnung der Dinge' (The order of Things) by Michel Foucault (1966), or 'Industrial Design heute', by Wilhelm Braun-Feldweg (1966), it, once again, shows the ethical force of interdisciplinary scientific work and high significance of questioning the relevance and presence of the knowledge sciences – beside offering 'open access formats' for all: Who owns knowledge?

Yet, aside from knowledge, behaviour also plays a significant role. The shift we see in design today signals a change in culture and materials, both of which have triggered a change in 'media behaviour' (see chapter 7): the power of a moral understanding of how the world works, as well as the technological revolutions and possibilities that have been developed since the end of the 1990s. This underscores my vision and method in design: reflecting upon design engineering, with the understanding gained through an open-minded cultural education, and comparing disciplines and linkages, **as well as** the characteristics of cultural behaviours that integrate and are respectful of others and otherness. The philosopher Paul Ricœur said the following about 'otherness': *'Neither prior perception nor a wish determine the focus of attention, but rather the naïve view, its innocence, the perception of the other as being the other. Through this active willingness I give myself into to the custody of the object. The true name for this attentiveness is not anticipation, but amazement;'*[3] (see Paul Ricœur, 1950, translation Huber-Fennell, 2021).

The design method mentioned above is being used in Germany and other European countries, relates more and more to the social issues being dealt within interdisciplinary and cross-cultural teams. **The aim is to combine design theory as a method for 'materialising immateriality', while using the events of the day as a trend, and placing trust in the naïve view of pre-school education in children and artists.** Sketching the needs of the future and building a strong foundation from the beginning – like it is demonstrated in the mind map: 'the Design Didactic Approach' – by expressing and communicating (mediating) your socially relevant ideas and problem-solving solutions. Let's sketch and design our future together while respecting otherness. (see graphic 05, p. 157)

2
Earley, C. and Ang, S., 2003, Cultural Intelligence – Individual Interactions Across Cultures, Stanford University Press p. 48.

3
Ricœur, Paul, 2016, (1950), Edited by Eßbach, Wolfgang et al, Das Willentliche und das Unwillentliche, Fink, Paderborn, translated by Louise Huber-Fennell, p. 191, (Original in French: 1950 und 1988, Le volontaire et l'involontaire, Philosophie de la volonté I, Flammarion), 'Weder die Vorwahrnehmung noch der Wunsch machen die Aufmerksamkeit aus, sondern die Naivität des Blicks, seine Unschuld, die Aufnahme des Anderen als Anderer. Durch diese aktive Bereitschaft begebe ich mich in die Obhut des Objekts. Der wahre Name der Aufmerksamkeit ist nicht Antizipation, sondern Erstaunen'.

Design engineering – and specialisations like 'textile engineering' – showcase complex problem-solving processes, which are currently being expressed by the shift in design and industry, that require us to look back in order to move forward. If you recognise the identity of the past, you will be more cognizant and encouraged for the future. Please be more aware of your imprint in relation to our planets benefit.

When I look back, I recognise that it was very important to go my own way. My imprint in the field of creativity in earlier times, as child at school, as an apprentice, as an entrepreneur and student, gave me the confidence to sketch ideas and visualise my thoughts, to discuss and evaluate my sketches and ideas with other people – reflecting, criticising, and also doubting whether I should create handcrafted or CAD-prototypes when my ideas became 3D models. Through my experience as a professor – lecturing and designing, reflecting and 'sketching out' – and by working on sustainable design engineering solutions with so many students, it has become obvious to me what the demands for the next design engineering education will be: If you have ideas and the possibility to experiment and express these through hand-drawn sketches – not necessarily with pencils – you can develop complex problem solving skills necessary for design concepts; and these will be the most important skills, an expert will need after the digitalisation. Through tangible experiences – using artistic techniques – and by reflecting upon objects and concepts of the past and present, you will be able to understand the needs that future designs must meet. If you can mediate your thoughts and ideas by sketching*, or expressing them through the use of materials, you will have the ability to use the working tools of Industry 4.0 – the design landscape of Augmented Reality (AR) included (see pictures, p. 223).

*Sketching in this sense refers to operating with the original function of 'design' by drawing, but also means to create the future in the sense of the German term 'Gestalten': understanding by thinking through drawing, generating, remembering and ideating and reflecting twice – not only by using connected 'Knowledge Banks'. 'Sketching' can be used in the metaphorical way of creating a future through ideas – materialising immateriality. This book will show you how. Do we have to call the next 'design engineer artist' as innovator of the future? as a very new model ...? or only a new name for an old 'form', (habitus) of 'industrial designer', thinking and designing in a holistic way, inspried by art, philosophy, human scientific educated...?

'And, in the same manner that tools steer and control the actions of the hand, the tools of the mind support and protect knowledge.'
Franz Bacon 1830 (1962 reprint, p. 26 in: 'Neues Organ der Wissenschaften'.)

Sketching the future together...
Marina-Elena Wachs, January, 2022

Graphic 01
Sketching the future together –
sketching in context
(see didactic approach later
in this book), M.-E. Wachs, 2021

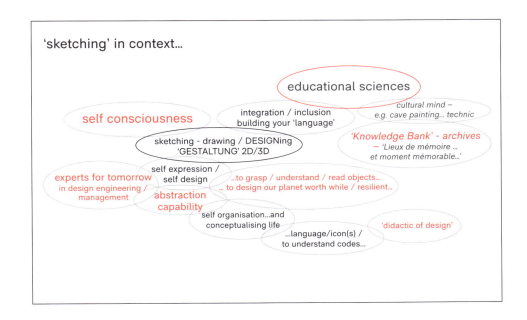

Sketching / Designing as Self-expression for imaging, mapping, reproducing, engineering,… to reflect, generate, ideate… design (of) the world – together!

(Wachs, 2022)

CHAPTER 01	12 — 23		Introduction
	14	1.1	Evaluating design heritage and design methods to discover new formats of design engineering in education and industries – is 'design theory' a method?
	20	1.2	Five ways to read this book
	21	1.3	Questioning the systems: education, creative and economic systems
CHAPTER 02	24 — 37		Industrial Design Engineering in the past: Industrial demands and design schools' heritages
	26	2.1.1	German industrial design made in Brunswick' – a less than ordinary and transdisciplinary approach and the question about training-worthy fields for industrial designers
	29	2.1.2	A historical perspective on the industrial design heritage of the Braunschweig University of Art – should it be considered 'design science'?
	34	2.1.3	Challenges: German and international provocations in Industrial Design
	36	2.2	A 'contract' for cultural education – the political willingness
IMAGES 01	38 — 75		Industrial Designer of HBK Braunschweig: Design success today

CHAPTER 03	76 — 101		The design term 'textile engineering', the significance of sustainable design drivers and the 'design shift'
	78	3.1	The driving forces of (textile) culture: 'Textile engineering' in history and German industrial design – a transdisciplinary look at working conditions, industrial changes and people's sources of identification (3.1.1. – 3.1.6)
	92	3.2	Driver of sustainable (industrial) design culture – the 'design shift'
	92	3.2.1	What does high-quality product design mean?
	98	3.2.2	The correlation between design and engineering
	99	3.2.3	The design turnaround, the digital revolution and de-industrialisation: How will the design shift impact design education?
PRELIMINARY APPRAISAL 01	102 — 103		Common and differing characteristics in product and textile design engineering – the business of 'industrial design' in the past and the 'design shift'
IMAGES 02	104 — 131		Textile design drivers', Textile Engineers' success today
CHAPTER 04	132 — 154		Education research in practise – comparing sustainable design studies and 'learning landscapes of togetherness' in Europe
	134	4.1	Collaborative non-hierarchical design engineering and management workshops – pan European workshops in Germany

	139	4.2	The method of 'material-based design thinking' with the aim of 'materializing immateriality' – a pan European workshop in Great Britain
	145	4.3	The European design engineering driving range – Learning lessons for a tangible, non-hierarchical education space towards material and immaterial togetherness (a collaboration between M.-E. Wachs, and A. Hall) (4.3.1 – 4.3.5)
PRELIMINARY APPRAISAL 02	155 — 157		The best Design Engineering models for Europe's future – providing resilient economic and educational benefits by 'forming' fine arts and making the future tangible
CHAPTER 05	158 — 173		Business models in design education and their entrepreneurial power – yesterday's future
	160	5.1	The entrepreneurial power of design schools – the past and future drivers of industry business models (5.1.1 – 5.1.4)
	170	5.2	Sustainable thinking in design education AND design engineering industry: the next academic and business models and the 'Management of Knowledge'
CHAPTER 06	174 — 207		Our aim to build and use cultural intelligence, using creativity to create ethical and holistical education landscapes
	176	6.1	The 'flow' in design – an extreme super power for working conditions or simply wishful thinking? (6.1.1 – 6.1.6)

| | 188 | 6.2 | Virtual Interactive Design (VID) – shaping digital and ethically correct spaces for creativity and its benefits to stakeholders (6.2.1 – 6.2.3) |
| | 197 | 6.3 | In consideration of a holistic design engineering education – integrated cultural intelligence (6.3.1 – 6.3.5) |

IMAGES 03	208 — 245		International creative perspectives – ideating future together International Guest Statements

CHAPTER 07	246 — 267		'Sketching' sustainable future together in a hands-on, non-hierarchical and holistic design engineering education culture – cherishing playgrounds for art and design engineering heritage by using 'Knowledge Banks'
	248	7.1	A position paper for interacting and integrating systems in sustainable design engineering and design education – state-of-the-art, holistic, non-hierarchical and interdisciplinary learning
	255	7.2	Rethinking design systems and codes
	261	7.3	E-Valuating European design memory by using a 'Knowledge Bank' – research responsibility in Ethics, E-technology, Education (the big 3 E's for cultural education and economy)
	265	7.4	Design Engineering – sustainable and holistic: education as 'industry' – a place for education and to remember!

EPILOGUE	268 — 269		
APPENDIX	270		Acknowledgement
	272		Imprint

CHAPTER 01

12 — 23
Introduction

1.1 Evaluating design heritage and design methods to discover new formats of design engineering in education and industries – is 'design theory' a method?

'There exist so many to adore, to love ... what a wonderful, fantastic creation is the world! How diverse! Diversity – that is nature's law. **Never you can play the piece of music twice in the identical way, with each note it will develop to another world.***'* (see Pablo Casals – Cellist) [4]

Sketching the future by 'materialising immateriality' means looking to design engineering in the past and education history. This allows us to revisit discussions from the past regarding the use of 'form and function' as a didactic design approach, while also focussing more on cultural memory and consideration of today's media behaviour. In turn, this gives us insight into the relevant creative connecting skills relevant to the future, and means, we have to question **design methods: Can they serve as 'the' working tools for the future if the present shift in design engineering is focussing on design theory as method? Also, is this what is required to make the world's design engineering industries and education programmes attractive and elevate their status? Design has to find the answers to more than just sustainability and digitalisation!**

'Creative connecting skills' are the answer for sketching, designing, to mediate and to gestalt the future' (Marina-Elena Wachs, 2019, Glasgow, EPDE conference).

In the past, we focussed on classical industrial design, and how 'form and function' [5] related to problem solving and practical elements with the highest possible aesthetical value (see Chapter 2/3). At the present time, it is more a question of social awareness in design solutions.

In other words, design concepts are being created with a stronger focus on generating social scientific and cultural anthropological relevance. Theoretical discussions are elaborated at the beginning of the creative process, and are highly valued parameters. This Systems Oriented Design process relates to a phenomenon this is occurring simultaneously: the change in the digital revolution.

As a result, the Era of Machinery is being upside-down: Up until now, economic enterprises – classical machine engineering – have represented status and success in the industry (see chapter 3). In fact, they have dominated the German economy for centuries and have been a dominant factor in the gross domestic product. However, in 2019/20 these powerful *German* industries have begun to experience a lack of demand for various reasons. Here are only a few:

---

[4] Casals, Pablo, in: ZEIT Klassik Edition, Buch 06, 2006, Zeitverlag Gerd Bucerius, p. 5: ‚Es gibt so viel zu bewundern und zu lieben ... Was für eine fantastische, wundervolle Schöpfung ist die Welt! Wie verschiedenartig! Verschiedenartigkeit – das ist das Gesetz der Natur. Niemals kann man dasselbe Werk zweimal genauso spielen, mit jeder Note wird es zu einer anderen Welt.', translated by Wachs, M.-E., 2020.

[5] See Rams, Dieter and Lovell, Sophie, 2009, Dieter Rams, Phaidon; ibid: Institutions like 'Rat for Formgebung – German Design Council; ibid: IF Design: Die Gute Form'; ibid: The red dot', Essen in Germany proclaiming and awarding each year the best design, related to history about 'form'. Today the categories changed.

- cheaper production by Asian and other Eastern competitors
- new product demands, resulting from a paradigm shift in technological production processes (e-mobility)
- the Internet of Things (IoT) developed a behaviourism that excluded products and objects, which mirrors the change in social behaviour today
- the Covid-19 pandemic crisis and its consequences around the world
- climate change.

In the end, the evaluation was able to show that these are all aspects directly related to the technological development and a **change in Western society's ethical values** (see chapter 4 and 5). This unique revolutionary situation is flanked by the global Fridays for Future protests and social strikes for democratic balance and other human rights movements for designing life. Overall, it can be seen as a challenge, but also a chance to focus on the framework of the creative education scene – particularly design – as well as how to coach the next generations on the skilful use of a 'Knowledge Bank' (Wachs, M.-E.) (see chapter 7), to help shape the future. The **term 'design' has to be seen as 'applied art'**, which must be kept in mind in relation to the value of art. *'In diesem Rundgang wird der Besucher zum Akteur, zu einem Betrachter, der sich sehen, sich bewegen, anderen begegnen sieht und dabei eine Umgebung erlebt, die er so noch nie gesehen hat. Zwischen diesen beiden Erfahrungswelten, einer Schule des Sehens und dem Jahrmarkt, wird sich die Lichtkunst des 21. Jahrhunderts vermutlich entwickeln.'* (Michael Schwarz, 2015).[6]

The interaction between disciplines and cultural codes resulting from the interplay between various industrial and institutional partners, provides great benefit in that it has changed how entrepreneurs get involved (see chapter 3, 5 and 6): these individuals interact, connect, trigger thoughts and actions, and **formulate questions regarding the future. But what are the 'right' questions?** You, the reader, are encouraged to explore the interview questions (see chapter 1.3) and the experts' answers throughout this book by the guest statements. (see chapter: IMAGES 01, 02 and 03)

In addition to a shift in the meaning of the term 'design', its status and industry partners have also undergone a change over the past 20 years: Aside from a **change in design methods** (see chapter 4, 6, 7), **an industrial change resulting from the fourth industrial revolution – the digital revolution is taking place** (see Schwab, K, 2016 and others) **in terms of the active roles of designers and consumers**. And, as we can see, the 'revolutionary power' is presently on the rise again – this is relevant for the necessary change. The following theses derive from of the paradigm shift:

---

[6] Schwarz, M., 2015, in: Otto, Julia et al, (Hg.), 'Scheinwerfer. Lichtkunst in Deutschland im 21. Jahrhundert', Kerber, Kunstmuseum Celle mit Sammlung Robert Simon.

1. Ethical questions have emerged within the last 20 years in 'design doing', and the process of reflecting on how to use design objects and concepts. **Therefore, design reflection and design theory can be considered as a design method.**

2. Human-centred design, design thinking, social design, integrative design, co-designing – to mention only a few design methods – are being applied more and more in the creative process. These methods are used to gestalt design objects or concepts at universities and in enterprises, to reveal the initial conception of social and environmentally friendly design solutions. These solutions have become less and less object-based in our capitalistic world and **we now have to 'materialize immateriality', in order to communicate contemporary, sustainable design solutions. Because understanding a process or a problem begins with replaying, imitating, simulating, and reflecting the world around you, your ideas and your inner world through 'sketches'. Nowadays, it is irrelevant, whether you 'sketch' with pencil on paper, with 3D materials or with light in augmented reality.**

In conclusion: Putting the **design object FIRST** is a thing of the distant past. **Subsequently, design methods were determined before the material or immaterial solution. However, even this approach reflects yesterday's conventional product design process.**

Today, the humanities are inspiring us and encouraging us to look deeper into the process of human cultural and medial behaviour as it applies to interdisciplinary interactive designing. In order to fully sense everyday design solutions, they must be conceived to improve the chances of more social acknowledgement and acceptance of good design – this is the task. Now, **the 'Knowledge Bank'** is expressed through the design process, more so than by design objects.

This may sound reminiscent of the Bauhaus, whose centenary year was celebrated at several congresses and exhibitions in 2019. Here we once again proved how innovative and impactful the structures of Bauhaus' education were on various industries.[7]

In 2018 and 2019, as a result of all the exhibitions and books published with more research findings that related to the centenary year, new questions have arisen about the impact the Bauhaus had at the time, and the consequences for industrially produced objects and design engineering education – which has lasted until today. This questioning is being approached in a fresh, new way (see chapter 5):

For example, at the end of 2018, the Anni Albers exhibition, presented in Dusseldorf and later on at Tate Modern in London, revealed two contrasting positions in terms of the exhibition concepts: The questioning of Anni Albers œuvre focussed more on the position of a 'textile designer' in Germany, whereas, in Great Britain, the focus was from a 'textile artist' point of view. European and global relationships were presented from a new perspective in those respective creative fields; e. g. with the conference 'Collecting Bauhaus' in Dessau in December 2019, and one important new *publication*, the book 'Bauhaus Goes West' by Alan Powers, published in 2019.

---

[7] See DHS congress Newcastle; Tate Britain London: expo: Bauhaus and Britain; V&A Museum London; Bröhan Museum Berlin expo: Nordic Design Die Antwort aufs Bauhaus; Dessau (Congress and different exhibitions 2019), *Dusseldorf, (Anni Albers, Kunstmuseum Dusseldorf),* Krefeld, to mention a few.

The Bauhaus life conditions and the dominant personalities – like Walter Gropius[8], Laszlo Moholy-Nagy, Marcel Breuer[9], and also the unrecognised women of the Bauhaus who were 'Beneath the Radar'[10] – all came into focus again in 2019:

As Powers underlines in 2019 *'As a German émigré, Margaret Lieschner was interned by the British during the war but succeeded in establishing a career as a designer for woven fabrics, often with specialised technical requirements for car and aircraft seating. As Professor of Weaving at the Royal College of Art, she influenced teaching for industrial design in textiles throughout Britain.'*[11]

In 2019, during the Bauhaus's centenary, other revolutions expressed socially relevant cries for changes in the way media is used to communicate: The Fridays for Future revolution, Extinction Rebellion activities, the French 'gilets jaune' demonstrations, and gender debates all stand in contradiction to the 4th industrial (digital) revolution, which is focussed on the power of 'technē'[12], expressing the wish for change. At the same time this is the 'ductus' of human beings: the self-expression of societies' feelings (Befindlichkeit): *'Ultimately, our technologies are separating us from the world as seen by Seneca and Aristoteles. The power they give us is no longer comparable to that which humans wielded in ancient times, or even at the beginning of the 20th century. Technology has fundamentally changed the way we perceive our responsibilities towards nature and future generations. It offers individuals new possibilities and feeds the desire to break through conventional barriers.'*[13] This philosophical view of Pelluchon is just as important as Klaus Schwab's economical perspective that questions how general working conditions will be – as well as those of the individual – once digitalisation as has become prolific (see chapter 3, 5 and 6).

Let us then take a deeper look at this scenario and how to formulate the 'right' critical questions – questions about society and the technical possibilities in 'industry' – that will help us move forward in the 2020s. Thereby, it should be noted that the terms 'industry', 'industrial' and 'industries' are currently in the process of changing, which should be taken into consideration, as explained later in this book. But let us focus here on what is different compared to the past, and what will influence the design engineering process in a holistic way – in terms of both education and 'industry'.

In consequence of the process of 'looking back at design engineering and education history' (see chapter 2 and 4), we can come to the following conclusions:

Economy and creative industries are showing more acceptance of each other. Furthermore, design is becoming more accepted, as is art: **Since the end of the 20th century there are more and more formats in art museums that aim to integrate design not only as a collection, but in a variety of design exhibition formats.** As mentioned before, this is thanks to certain personalities who think in terms of 'art + design' – influential people, entrepreneurs and trailblazers, like Gropius (as architect and director) – who have functioned as significant triggers. Other examples include: Paola Antonelli (former curator at the Cooper Hewitt Museum and now at MOMA New York), or Daniel Claude Hug (director of Cologne Fine Art Fair the format 'Cologne Fine Art & Design') and like Michael Schwarz (former

---

[8]
See MacCarthy, Fiona, 2019, Walter Gropius Visionary Founder of Bauhaus; ibid: Powers, A., 2019, Bauhaus goes West,; ibid: Rössler, P. und Otto, E., 2019, Frauen am Bauhaus – Wegweisende Künstlerinnen am Bauhaus, Knesebeck.

[9]
See Shipley Art Gallery, Newcastle.

[10]
Powers, A., 2019, Bauhaus Goes West, Thames & Hudson, p. 166 ff.

[11]
Powers, A., 2019, Bauhaus Goes West, Thames & Hudson, p. 144 ff, table 18.

[12]
Note: the Greek term 'technē' is meant here.

[13]
See Pelluchon, C., 2019, Ethik der Wertschätzung – Tugenden für eine ungewisses Welt, wbg Academic, p. 29 f, (translated by Louise Huber-Fennell); (Pelluchon, C., 2018, Éthique de la considération, Edition du Seuil.)

professor and president of University of Fine Arts Braunschweig), with open minded, unconventional habits and decisions pro 'art + applied arts' to mention only a small selection of *forerunner*.

Also, very impressive happenings, formats like the Venice Biennale and other temporary exhibitions, are serving as catalysts for developing new strategies. The ever-changing leading curators highlight different perspectives and are urged to implement innovative formats. Another important place to be, is the annual performance of innovation: Salone del Mobile at Milano, the most creative incubation tank for art, design, and lifestyle in a holistic, creative sense. This was proven again in 2019, by the 'Triennale' events and the catalogue by Paola Antonelli, addressing the subject of sustainability, for example.[14]

Perhaps all these drivers are pointing to a 'new interdisciplinary ordinariness' and will guide us through the business of art, design and the creative industries + sciences in the future. This could ultimately lead to more highly valued objects, as well as better work, living and education conditions for all people.

This postdoctoral research study is an offer to open your mind to more 'European and globally integrated togetherness' in design engineering, as well as interlinked non-hierarchical interdisciplinary activities. It showcases the advantages of providing cultural education, from kindergarten-age all the way to retired experts + practical projects. This **'position paper'** (see chapter 7) **presents** a modified 'designing format for the future', **and provides a variety of collaborative pathways for attaining sustainable design solutions – a selection of the most important elements are:**

- design engineering in cross-cultural, interconnected digital and analogue working spaces
- designing by means of cross-generational courses that also include retired experts (vocational education) and children
- engineering the future through artistic education – the significance of 'sketching the world'
- encouraging children at pre-school age to take responsibility
- implementing non-hierarchical learning landscapes
- designing in pre-schools through play – please take this task seriously
- acknowledging the impact of art and culture in everyday solutions
- including sustainability and design engineering in school curricula (in Germany)
- intelligent and interdisciplinary connected thinking and creating in Mixed-Reality projects – but this requires a knowledge base
- using Artificial Intelligence (AI) and Emotional Intelligence (EQ) to promote a new attitude in 'Integrated Cultural Identity' (ICI)
- using design theory as a method – materiality versus immateriality
- design engineering and 'textile engineering' must begin as a playful activity in pre-school
- sketching by hand and making future realities tangible by drawing – thinking through drawing – this forms the basis for sketching in Augmented Reality (AR).

[14] See Antonelli, P. and Tannir, A., 2019, Broken Nature, XXII Triennale di Milano, Mondadori Electa S.p.A., Milano.

Besides the 'must haves' for the future of Sustainable Engineering Design (SED), this research study gives you insights into the paradigm shift regarding the use of design concepts and the objects produced. The ongoing change in cultural behaviour – and indeed of 'media behaviour'[15] – is extremely relevant to this process.

This changing behaviour, makes the need for a concomitant acceptance of design in the 'creative business landscape' quite obvious: In the past, design was regarded as an object, a historical artifact (a collector has a passion for classical furniture, for example). The focus was always on evaluating form and function, the high-quality aesthetics and long-lasting design language. The pinnacle of such endeavours is demonstrated by becoming a highly esteemed member of the hall of fame in museum collections (e.g. MAY DAY lamp by Konstantin Grciç at MOMA, New York).

Today, design is part of the paradigm shift and must be accepted as part of society in Europe. This means not only viewing design as proof of culture and cultural heritage, but also as a way to strategically manage social responsibility, as well as have a positive influence on everybody's daily habits and environments.

This book, 'Design Engineering – sustainable and holistic', **encourages you to view design history as yesterday's future of design engineering and maintain a 'naïve view' towards sketching the future.**

---

[15] Note: 'media behaviour' is a new term coined in this research study, and relates to design theoretical point of view of 'material mind', as defined by Marina-E. Wachs in 2007, to the anthropological point of view of 'material behaviour', as defined by Michal B. Schiffer and applied by Hans Peter Hahn.

1.2 Five ways to read this book

The idea of so-called 'skin in the game'[16], is a phrase coined by the entrepreneur Affentranger, who claims that designers need to fully invest: their knowledge, creativity, time AND their own capital. On the one hand, this can be interpreted from an economic perspective, in other words taking responsibility by investing private capital for all entrepreneurial affairs. On the other hand, this now also applies on the meta level and could thus be interpreted as a further development of the 'lean in' strategy by Sheryl Sandberg (2013), which refers to the metaphorical interpretation of taking responsibility for your actions, and your inventions. Having 'skin in the game' is more in demand and is more valuable than ever before. **Being motivated to act, and show dedication for enabling change in a *corporeal* way is what is needed today.**

This phenomenon is expanded upon in chapter 7, but based on this premise, you can benefit from this book, by reading it in a variety of ways. You can take these five paths to focus your aim on this postdoctoral thesis about 'Design theory and interdisciplinary practise in design engineering and education in history, the present and the future':

— You may first answer the interview questions (see the following list of questions from 'December talks 2019' by Wachs). Consider your thoughts as presented from your own personal point of view as well as a business perspective, to formulate the needs of design engineering processes in the future.
— You can use these questions to generate sustainable solutions in design concepts and for design education.
— You can discover how the structure's common thread provides understanding: it guides you through a cultural – sometimes more anthropological – holistic, interdisciplinary view, yet remains *based on industrial design and design engineering.*
— You can follow the story of design education history as it relates to the development of design methods, and you can generally dig deeper into innovative advanced methods for your business – whether you are an (industrial) design nerd, driver of innovation, or an expert from another branch.
— You can enjoy the idea of a life that is designed better, or you can go to pre-school to play designing with our future generation of designers: This is always a gift and gives you quality time for sharing experiences and ideas with the best naïve minds we could wish for.

[16] Affentranger, Anton, 2019, Baustellen, Innensichten eines Unternehmens, Münster-Verlag; *please note: the term 'skin in the game' by Nassim N. Taleb in 2017.*

1.3 Questioning the system: education, creative and economic systems

The following questions relate to the relationships between the elements and the pre-requisites of **Design – Art – Design Engineering – Creative Industries**, in terms of creating sustainable solutions. With the help of the questions, you may get a first impression, or develop some key thoughts for yourself. These questions were part of the 'December Talks 2019' by Marina-Elena Wachs, that were held in Germany and Great Britain.

1. **How great of an impact do art and design have on economic structures and economic benefit in Europe, and how are design and art influencing the circular economy?**

2. What **lessons can we learn from art and applied art in history?** How would you describe particular benefits?

3. During an interview with Inga Griese, for the magazine 'Welt am Sonntag – ICON' in October 2019[17], the director of the V & A Museum, London, Dr. Tristram Hunt was quoted as saying the following about the impact on design history: **The founder of the museum, Prince Albert, would pay more attention to 'design education' at school today.** What do you think about this as an **educational strategy**?

4. What do you think about the **currently increasing acceptance of design**? 100 years after the Bauhaus, what are the driving factors leading to more and more respect for design as objects and design as a discipline, in spite of the fact that it has been the 'little brother' of sciences of art for a long time?

5. In my opinion, **particular people and museums have been especially innovative in this field of thought**; e.g. Paola Antonelli and the Cooper Hewitt Museum in earlier times, and today's curator at the MOMA in NY; the director of Cologne Fair Fine Arts, Daniel Claude Hug, for the 'Cologne Fine Art & Design' fair. Other important innovative drivers that are focal points include, for example, art and design at 'Salone del Mobile – Milano', the Victoria & Albert Museum in London, the Venice Biennale, the MOMA in New York, and the MOCCA in Cape Town. In this field of cultural involvement, another point of view is presented by individuals initiating museums for society, like the Frida Burda Museum or the Zentrum für Kunst und Medien (ZKM) Karlsruhe. Also, the Kunstmuseum Wolfsburg, whose primary aim is to draw in the regional population from the 'working city' of Wolfsburg, here, the foxus lies on the impact of cultural education.

---

[17] See Hunt, Tristram, 2019, in: ICON October I 2019, supplement 'Welt am Sonntag', Issue, p. 78 f.

Furthermore, an increased commitment to having a positive social impact has been demonstrated by entrepreneurs, such as Erck Rickmers for example. He founded the 'Humanities and Social Change – International Foundation' in 2018 and opened the 'The New Institute' in Hamburg, Germany, in 2021.
What is the reason for this social commitment? Is it truly social (and political) responsibility and commitment? Or a ploy to gain political influence? Or, is it perhaps a result of the government's lack of commitment towards cultural educational institutions?

6. Within the **predominantly Anglo-Saxon discipline of anthropology, design as an applied art, in particular 'design engineering' as a discipline, was more accepted in Great Britain than in Germany, in the past:** for example, by respecting 'cultural behaviour' (see Küchler S.) and 'material behaviour' (see Schiffer, M. B.) – a perspective that demonstrates a difference when compared to the behaviour in Germany. **This fact leads me to the next question: What differences can be identified between cultural behaviours and the value placed on cultural heritage** and handcrafted techniques based on their respective past? Do you have any experiences with which you could compare the characteristically British or Scandinavian use of objects, over the course of history? And, **do we need a more articulated acceptance of 'design' and acknowledgement of its relevance in school curricula?** (– Please to consider the fact that young teens can already select design engineering courses at school in Great Britain.)

7. **Where have we come from – where are we going?** These are the key questions, but they seem to be taken less seriously in Germany: **drawing benefits from the past** is considered 'old school', and is viewed derogatorily.
If we compare this to the Comité Colbert at France, and many Italian or European Fashion Foundations (e.g. Armani Silos, Prada Foundation, Fondation D'Entreprise HERMÈS), we see a great contrast. They offer educational programmes – not only in museums for the public, but also for academic institutions with the support of entrepreneurs (e.g. the house of Fendi or Brunello Cuccinello, **as well as others in the Mediterranean region). I am convinced this is the best way to secure the knowledge for the future, thus creating a 'Knowledge Bank' (Wachs) that functions as an 'archive of knowledge', from which future generations can profit**. What will happen if the German process continues to show less acceptance for the use of this heritage – expressed through lack of financial support and less awareness of Germany's (textile) **design** (engineering) heritage?

8. Let us take a look to the future developments in art, what do you think about that? For example, if you look at the **Biennale in 2019**. This was a great social and political statement in the form of art, which was temporarily presented **at the Canadian pavilion, for example: a group of Inuits had been selected to raise awareness for all forms of climate change on earth.**

Does this mean that art will have the function of raising awareness for all social/political issues in the future – in creating such of a kind documentary scenario, for example? And will 'simply' making people aware of a social or environmental problem be enough?

9. Have you identified any parallel phenomena in design that are less object-based than ever before, and that there is an increasing focus on social design and integrative design, to solve society's problems? **And, how great an impact will the act of reflecting on art and design have on the post-digital society? – or what do you think about the initiatives of residence programmes of CERN and others, supporting an experimenting melting pot of art + technic, of design engineering – art – sciences?**

10. In my opinion, **the most important capital and resource we have in Germany and Europe are the humans**. Thus, it follows that cultural education, knowledge management, and education management in a broader sense than 'controlling' needs to be promoted. Do you agree? **How would you invest in cultural education, and who should invest?**

11. Would you like to add a personal statement? Think about the future of cultural education, the next era of cultural heritage, or about the future of your business – and now **make a wish**.

      With these questions regarding the systems (education, creative, economic systems), you are encouraged to find answers within the following chapters. Let us start by looking at design engineering in the past, and focussing on the comparison between German and British industrial design education heritage and places to remember…
      Enjoy and never stop thinking or asking questions in design engineering – put all your heart, soul and mind into the paradigm shift.

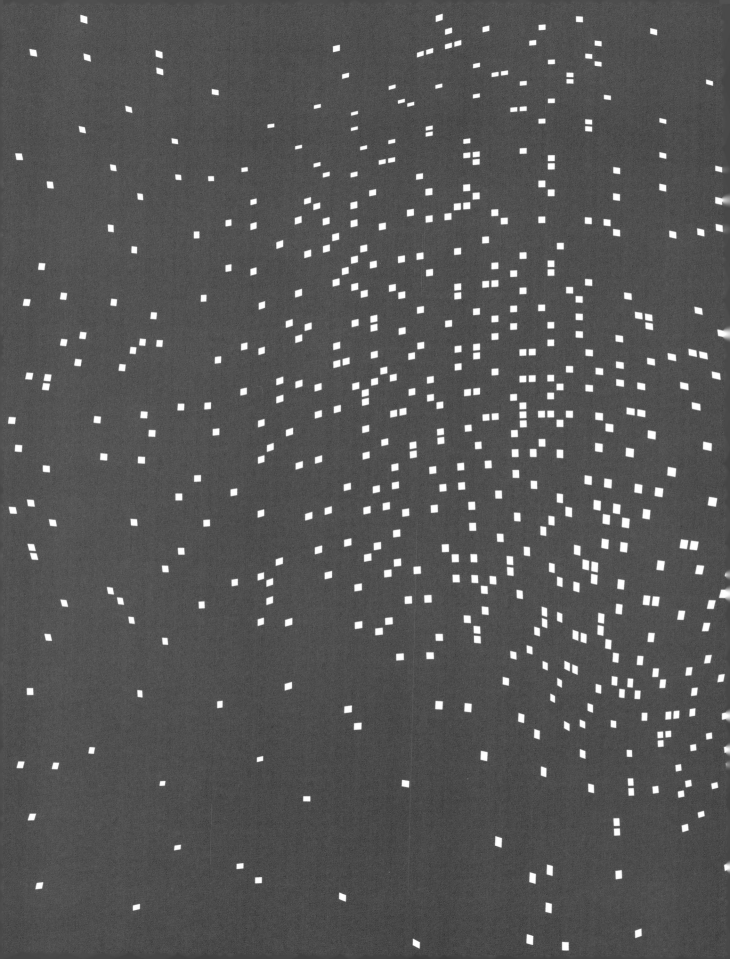

CHAPTER 02

24 — 37
Industrial Design Engineering
in the past: Industrial demands
and design schools' heritages

2.1.1 'German industrial design made in Brunswick' – a less than ordinary and transdisciplinary approach and the question about training-worthy fields for industrial designers

Following, you will find an English summary of a former German text written by Marina-Elena Wachs, in which you can find exceptional historically significant research results – of a case study about Brunswick, a former 'Werkkunstschule' (design school) like 20 others in Germany's past – that could be beneficial for 'remembering your archives' in Europe. It may well be essential to dig deeper into the original cultural language code, particularly in German, to create, for example, an understandable industrial design foundation in the future that the next generation can profit from. The 40-page manuscript by Wachs, in German text, provides sources of the former 3-dimensional archive of the Braunschweig University of Art. This archive could give deeper insights into Germany's impressive industrial design heritage. Ergo, investments in archives serves the idea of creating 'places to remember'. This will be explained more in detail in the successive chapters, and additionally by case studies.

Please follow by reading chapter 2.1 to 2.3 and the ongoing comparative studies, and subsequently their significance of other European design engineering education models found in chapters 5, 6 and 7. This will help you 'sketch the future' of industrial engineering.

Both at the beginning, as well as at the end of the review of industrial design education in the past, we will come to the question: What is design-worthy – 'less than ordinary' – in industrial culture in terms of industry and market demands, as well as the needs in education systems and their paradigm shift' we regard today.

The Braunschweig University of Arts has a long-standing reputation as a distinguished university for fine arts, although design has sometimes enjoyed the position of a 'little brother'. There have been a few times when the reputation of industrial design at the Braunschweig University of Arts was equally as strong as that of fine arts.[18] In order to dock onto[19] the existing trans- and interdisciplinary forces and most of these university's history (like for example the Braunschweig University of Art), one possible path could be: Imagine 'industrial design' being represented to the world via diverse profiling fields – as a kind of an own 'brand'. That means communication that is in line with international standards: one profile could be to claim the phrase 'Industrial design made in Brunswick, Germany', for example. It can be said that all of the author's 'visions' can be transferred to other schools for Design Engineering. Within this scenario, possible

18
See for example, times under Prof. Schürer's leadership when the reputation in the industry was very good, which motivated students to want to study at this university. See interview protocols by M.-E. Wachs with alumni, students and professors – German text., 2013.

19
See Helga Nowotny – as a cultural scientist – in her 2005 book 'Insatiable Curiosity Innovations in a Fragile Future' shows that one has to build on previous structures in order not to take the community with you in an effective way.

visionary aspects could be raised for discussion, to generate potential for a design theory-based and reflective design training. We will come back to this point of view in chapter 2.1.3.

But first take a look at the past – a look at the cultural heritage of industrial design at the Braunschweig University of Art: If we dare to view the entire university as one, which integrates the fields of art, design and sciences – we can declare that this is part of the 'Braunschweiger Model', named by Michael Schwarz, (see Braunschweig University of Art' module manual, 08/2013)[20]. That can be also proven with the University's subsequent 'yearbooks', which were created under the management of President Michael Schwarz and President Barbara Straka[21].

The research outlined in this chapter is based on research work completed in 2013[22], within the archive of Braunschweig University of Art, and finalised with comparative studies in 2019. This research revealed that surprisingly common historical aspects were elaborated – such as interdisciplinary working and textiles as design worthiness, for example – when comparing the Braunschweig University of Art to other design schools (e.g. to the Hochschule Niederrhein – University of Applied Sciences).

We must start with Karl Wollermann, the first director of the Braunschweig University in 1963, who led multiple textile classes, and was able to elevate fashion design to being a worthy field of study for a university of applied art. Wollermann's efforts could be interpreted as being design-theoretical, which evolved into a more product-semantic approach during the 1960s. Later on, he focussed on the use of textiles as material in the design and architecture of the 1990s. He went on to utilise technical textiles in boat and automotive constructions (Please compare representative designs by Braunschweig University of Art Alumni (see IMAGES 01)). When comparing these different views of design disciplines, the question of design worthiness becomes obvious, as well as the awareness that **there is nothing more emotional than fashion design AND automotive design!**

In the year 1953, Prof. Wollerman was the first director to run the institution as a university[23], after having restructured the preceding institution called the 'Deutsche Werkkunstschule' (German Design School), of which only 20 existed in Germany. The 'Deutsche Werkkunstschule' in both Braunschweig and Hannover derived from the previous Master School of Craftsmanship (Meisterschulen für das gestaltende Handwerk). The approach of these 20 German Werkkunstschulen can be summed up as follows: they 'see their education model as a continuation of the tradition of the Kunstgewerbeschulen and the Deutsche Werkbund'[24].

Let us look back at the industrial design heritage with German example of Braunschweig University of Art – its aspiration for beauty, functionality, interdisciplinarity and sustainability, with dominantly 'hard' male design, but also some fine, very valuable female design concepts. Look with appreciation at regional strengths, while showing consideration for comparable international perspectives. After reviewing alumni designs and researching the historical background of the University's development, one 'model' could be sketched – visionary by the author – to represent an advanced university of art, in which the disciplines – art, design, and sciences – serve as three pillars (like the 'Braunschweiger Model'):

---

[20] Please note: The Module Handbook of the industrial design department was being revised at the time, see interview with E.K., 13.02.13; ibid: Industrial Design Modulhandbuch der HBK BS, http://www.hbk-bs.de/imperia/md/content/hbk/hbk/studiengaenge/industrialdesign/modulkatalog_ba_industrial_design_gesamt.pdf., online: 08.08.2013 and 03.12.2020.

[21] See Straka, B., 2007, Die Freigeistige unter den Kunsthochschulen, in: Straka, B., 2007, HBK Katalog Kunst Design Wissenschaft, Volume 1-2, p. 10.

[22] Note: upon invitation by Hannes Böhringer and Dr. Claudia Bei der Wieden, supported by Michael Schwarz.

[23] See von Amelunxen, Hubert + Bei der Wieden, Claudia, HBK BS (Hg.): 50 Jahre HBK Braunschweig, Geschichte einer Kunsthochschule und ihrer Vorgängereinrichtungen, Hbk Braunschweig, p. 192; (see commemorative publication: 'Werkkunstschule Braunschweig in 1953')

[24] Bei der Wieden, C., in: von Amelunxen, H. + Bei der Wieden, C., 2013 (Hg.): 50 Jahre HBK Braunschweig, Geschichte einer Kunsthochschule und ihrer Vorgängereinrichtungen, HBK Braunschweig, p. 191, translation: Louise Huber-Fennell, 2021.

This model can be characterised through design theoretical and design scientific themes as postdoctoral theses (Habilitationsschrift), even more design practice with the help of European and international trans-university collaborations, satellite studios and even more master specialisations or profiles offered by multi-lingual, cross-cultural schools. The aim could be a stronger incorporation of synaesthesia and a much earlier educational reciprocation between all levels of schooling.[25]

Please also take a look at the 'incite design' in the 'excursus images', but we will examine the Braunschweiger Model: German industrial design made in Brunswick – a less than ordinary and transdisciplinary approach. Because at the end, we have questions and understand the power of our design education heritage, and how to bring the next designer generation into focus.

---

[25] Note: this is the author's intention and the advice to the Braunschweig University of Art (some of these ideas were already provided by the author in 2013). These visions are transferable to other industrial design universities.

2.1.2 A historical perspective on the industrial design heritage of the Braunschweig University of Art – should it be considered 'design science'?

Will practise-based industrial design education transform itself, from a design theory-based education to design science? Answering this question requires a retrospective look at yesterday's design engineering. We can begin in the 1960s, when the Braunschweig University of Art was founded, and Roland Barthes' study of the fashion system (original: 'La système de la mode') also attempted to find and generate a 'theory of design'. Barthes, as a philosopher, said:

'My main intention has been to reconstitute step by step a system of meaning, [...] to reconstitute the semantics of actual Fashion. [...] The object of analysis [...] is a true code, even though it is always only 'spoken'. Hence, this study actually addresses neither clothing nor language but the 'translation', so to speak, of one into the other, insofar as the former is already a system of signs.'[26] In 2022, we can take into consideration that Barthes' system referred to a system of signs, of cultural codes and behaviour.

In this context it is important to understand the difference between the power of product design within **a three-dimensional object** and fashion as a representative medium, **which creates a design system**. This translates into a design system with typical characteristics, and a design code for an object – much like how philosophers in ancient times described the Roman way of clothing, or how the 'New Look' in 1955 gave design new symbolic meaning. Imagine a field study of 'the black box' in product design compared to the 'Little Black Dress (LBD)' in fashion design. This comparative field study provides us new ways of thinking about a design system:

If we view the 'Little Black Dress' as a fashion design classic and compare this artifact, with the 'black box' in product design, you have to know how to read (into) the artifacts and reflect the semantic meaning within each. There are different aspects and methods used to look at design systems within their historical framework – e.g. from a psychological perspective, pertaining to a sociological agenda, or semantically when analysing the meaning of artificial codes. In addition, there are a variety of ethnological perspective of human ways of 'using' things, sculptures, services and concepts, and has to put into context. Otherwise, it is not possible to interpret artifacts correctly.

[26] Barthes, Roland, 1983 (1967), The fashion system, translated form the french by Ward, Metthew and Howard, Richard, London: Vintage Books, p. x / foreword f.; Compare: French original: La Système de la Mode, 1967, p. 8: 'Ce qu' on a voulu avant tout, c'est reconstituer pas à pas un système de sens [...]'.

As a design theorist based in the education of crafts and industrial design, the author (Wachs) has taken another analytical path for generating a valuable system of design codes: 'from cut to context and back – a six-level analysis on how to read and create things and concepts.' In this chapter about industrial design – yesterday and today – using this comparative model in fashion theory allows us to take design theory applied in industrial design at transdisciplinary occasions, to show similarities. The analytic system has been adapted, based on Wolfgang Kemp's model of the theory of architecture[27] (2009). We may well benefit from looking at design history to find a sustainable model for designing the future. At the same time, the semantic meanings of artifacts are as equally important as the act of creation. – Insofar 'thinking about designing' in creation and the method of 'thinking by drawing' shows how significant this fact is, on which will be elaborated in the following chapters.

In this chapter, you as the reader, are called to truly sense the change of perspective that we are currently experiencing in design. Also, this book requires somewhat flexible thinking, and for you to jump between perception levels. We can also take an inventory of the past 50 years (1963-2013) of industrial design at the Braunschweig University of Art, thanks to the physical archive made available in 2013. This archive allows various research topics to be deepened later on. Please take the Braunschweig University of Art as a model for the predominant Zeitgeist of design education in German design during those. This allows a comparison to be made with other universities, such as HfG of Offenbach am Main – University of Art and Design, or the Weimar School of Design, where design training is performed with smaller or larger profile variations, in creative design study programs in Germany.

'At this school, the handcrafted work is to be given the nobility of a genuine form culture, which must first assert itself in the Brunswick and Lower Saxony areas, in such a way that together with other art schools form values are created, which later transform into economic values.' (Karl Wollermann, 1953, translated by Wachs, 2020).[28] These words spoken by the first director of the 'SHfBK' (former Name of the Braunschweig University of Art), Karl Wollerman, draw the link between the former institution[29] and the present University of Art in Lower Saxony. Furthermore, as provocatively cited from 1953, this statement once again reflects the current perspective, and also points out the comparisons between fashion theory, design theory or product semantics, in an effort to imply design worthiness: What topics and problems have teachers of industrial design deemed worked enough to work on, maintain and assess in the past, and how does it compare to today? What makes the Braunschweig University of Art's industrial design history so impressive that fashion and furniture design seem to have given way to the tangible, powerful industrial design at the University, since the 1980s?

First, industrial design played a dominant role in the design of machines and 'human – machines – interfaces' were the focus, and the university's credo 'industrial good design', which represented the approach to solving design problems in the 1980s and 1990s. Nevertheless, the trend towards more 'process managed designing' and a theoretical

---

[27] See Kemp, Wolfgang, 2009, Architektur analysieren – Eine Einführung in acht Kapiteln, Schirmer / Mosel.

[28] Wollermann, Karl, 1953, Werkkunstschule Braunschweig im Jahre 1953, in: Festschrift zur Eröffnungsfeier des neuen Gebäudes der Werkkunstschule Braunschweig und anlässlich ihres einhundertjährigen Bestehens, 18. April 1953, hergestellt in den Werkstätten der Werkkunstschule Braunschweig: „Das Werk der Hand soll an dieser Schule den Adel echter Formkultur erhalten, die zunächst im Raum Braunschweigs und Niedersachsens sich so zu behaupten hat, daß in Gemeinsamkeit mit den anderen Werkkunstschulen Formwerte geschaffen werden, die sich später in Wirtschaftswerte verwandeln.'

[29] See Bei der Wieden, C., in: von Amelunxen, H. + Bei der Wieden, C. für HBK BS (Hg.): 50 Jahre HBK Braunschweig, Geschichte einer Kunsthochschule und ihrer Vorgängereinrichtungen, HBK Braunschweig.

concept-based approach, was evident in several courses. Second, the male teachers had a particularly strong influence on political and labour market-related matters at that time.

At the university, fashion design solutions were on demand at that time, for example in the form of illustrations in the parallel study program Graphic Design. In this respect, fashion and furniture seem to be tolerated as 'excursions' in the 1990s, both in project and study work. However, it is surprising that, due to the great interdisciplinary nature of the University, former 'Graphic Design' and 'Industrial Design' programmes were combined into one department in the 1970s, under the concept of 'Gestaltung'. In Graphic Design, the 'study focus was on fashion graphics/stage costume', but this was closed in the year 1978. In retrospect, Karl Wollermann, as quoted at the beginning, functioned as a mediator: He was an architect, who began his Brunswick days, as director of the former school of art, and served later as a professor of textile art, after which he became a leading position of textiles, weaving and embroidery in the textile classes in 1963 until 1967. This development consequent related to the outstanding created class for weaving and stitching in the year 1956, under the guidance of Prof. Karl Wollermann and Gudrun Dünnebier responsible for the laboratories, as stated by Bei der Wieden in 2013.[30]

'It was a happy coincidence that professor Karl Wollermann, who was open to the textile arts, was established in Braunschweig in 1951, the centre of this area, which is so important for the history of image embroidery in Germany, took over the management of the textile class in addition to the directorship of the Werkkunstschule and began to continue the legacy of the past.'[31] (see Bert Bilzer, Director of the Braunschweig Municipal Museum 1953-1977). Textile art – then still considered a female domain – can certainly be described as a niche for Wollermann, as well as a unique selling point for his artistic work. (see Wollermann's stiching motives of vases, HBK Braunschweig archive, 2013)

From this point of view, it is surprising, if we consider the time frame and the different social conditions and role models who were working and studying in this field – after all, looking at the number of men, who were working in 'female design' fields. It was a time when, for example Wilhelm Braun-Feldweg – a designer and professor for design in Berlin – demonstrated the male position in his book 'Industrial design heute' in 1966. Braun-Feldweg stated: *'Furthermore, several men (architects, craftsmen, engineers, painters and sculptors) set to work with their hands to give the so-called industrial production and our environment a new and unique look.'*[32] Thus, he proclaimed that the new term 'industrial design', also used in Germany (and at Braunschweig University of Art) at that time, to express a modern understanding and mindset for new study programmes.
By the way, the very well-known fashion designer Wolfgang Joop studied during the 1960s at Braunschweig University of Art in the department of fine arts (see Joop, W., et al, 2013, UNDRESSED).

According to Braun-Feldweg, the term 'industrial Design' was taken because the Anglo-Saxon countries had used it earlier in established fields, and because the German word 'industrielle Formgebung' or 'Formgestaltung' seemed to be too complicated in the international context.

30
See Bei der Wieden, C., in: von Amelunxen, H. + Bei der Wieden, C., 2013 (Hg.): 50 Jahre HBK Braunschweig, Geschichte einer Kunsthochschule und ihrer Vorgängereinrichtungen, HBK Braunschweig, p. 194.

31
Bilzer, Bert, 1961, DIE KUNST und das schöne Heim, Sonderdruck, München: Verlag F. Bruckmann, p. 112, translation by Wachs. M.-E. 2020.

32
See Braun-Feldweg, Wilhelm, 1966, Industrial Design heute, Rowohlt, Reinbeck, p. 186, english translation by Wachs, M.-E. and Huber-Fennell, L.. 2020.

But it is obvious that the Werkkunstschule's newly claimed profile seemed to promote 'industrial design' more than 'textile classes' at that time. This unique selling point combined with the integration of more women, appears to have been a strategic decision of Wollermann and Co. However, Karl Wollermann (and Bodo Kampmann) were announced as part of the jury to the 'if design award 1945', it underlines Wollermann's strategical position in leading the university with a more holistic point of view. This can also be underlined by his engagement for a class of drawing for children and young people.[33]

At the same time, Wollermann's aim points to the combination of disciplines, aesthetic requirements and networking activities of the study programmes. This was designed to form values both within and outside the university: (his words in the commemorative publication for the opening of the new building of the Werkkunstschule Braunschweig on the occasion of its 100th anniversary in 1953), he points out that shaping culture through the 'value of form that is transformed into economic value'. These values are generated by means of industrial design, which still holds true to this day. In the remaining text of this opening speech, he made it clear that the symbiosis between design (at the Braunschweig University of Art) and architecture (at the Technical University of Braunschweig) has continued through history since 1953.

In addition, it shows how important a symbiosis of collections could be, which perhaps, can be proven by the 'Form Collection of Walter and Thomas Dexel'. Dr. Thomas Dexel worked as a Professor of Art History at the SHfBK (former name of Braunschweig University of Arts) from 1963 to 1980, and also as a librarian at the preceding Werkkunstschule. The 'form collection of Walter and Thomas Dexel', was initially located at SHfBK Braunschweig (1963-1980), and was then exhibited in the Villa Gerloff in Braunschweig (1983-2003). Later, the collection was in the possession of the Municipal Museum, in Braunschweig, and was subsequently hosted by the Städtisches Musem am Löwenwall, in Braunschweig. This collection can – and should – be viewed as one of Brunswick's fundamental collections of 'form(s) theory', and thus not solely interesting for innovative and cooperative industrial design projects in the future. Wolf Karnagel's (see Wolf Karnagel as alumnus at p. 40) collection can be viewed as another opportunity to look back at 'yesterday', in order to shape the future. This process represents the method of 'design theory in practise'. This archive can be used not only to teach new generations to remember the 'lieux de mémoire' (Nora, P.) – places to experience the past – but also to gain respect for the value of the past innovators in design history and engineering. – Beside this fact, it can give interesting insights in the 'provenience research in design' (Wachs, M.-E.), which can be claimed as a research subject of tomorrow. – This allows students to benefit from the cultural education and design engineering experiences in a condensed manner before and after their studies with holistic view.

In the industrial design department at the Braunschweig University of Art, between the 1990s and 2000, they figured out that the goal was to obtain methodological skills for self-learning, to be able to adapt to a 'flexible' world – the creative industry was not only producing industrial

---

[33] See Bei der Wieden, C., in: von Amelunxen, H. + Bei der Wieden, C., HBK BS (Hg.): 50 Jahre HBK Braunschweig, Geschichte einer Kunsthochschule und ihrer Vorgängereinrichtungen, p. 193 et al.

design goods and serving the demands of industry 2.0. The generation of industrial design students at the end of 1990s placed a high value on quality, and profited from the academic resource, consisting of design reflection and cultural education, based on design theory and scientific cross-pollination (for example, with psychology, sociology, design methods, semantic relationships of form, typography and symbolic meanings of signs). They continue to value these resources today (see IMAGES 01 and IMAGES 03). The students were prepared for the future as 'design problem solvers', and were able to reduce the world's complexity (see Norman, Donald, Living with complexity; ibid: Simon, Herbert, The Sciences of the Artificial; ibid: Sennett, Richard, The corrosion of the character). They had been trained by Prof. Dr. habil Holger van den Boom – particularly on design theory and his ground breaking, initiated publication, to define the field of design sciences in: 'Öffnungszeiten' –, and Prof. Dr. Bernd Löbach – particularly on the subject of sustainability – and by others, like Felicidad Romero-Tejedor.

In 2000, Prof. Dr. Romero-Tejedor discussed the role of design education, and stated: *'Obviously, with the help of design education, it is not only about serving the industry in present times, but to create a general horizon, that mediates enough autonomy of oneself, for being prepared for tomorrow's demand.'*[34]

Please see in the following best-practise models of former industrial design students, **who attended the Braunschweig University of Art and are from various study generations, and work as designers in a range of disciplines today.** (see IMAGES 01, p. 38 – 75)

---

[34] See Romero-Tejedor, F., 2000, Design education, Ihre Rolle in der Zukunft, in: Van den Boom, H. (editor), 2000, Entwerfen – Jahrbuch 04, HBK Universiy of Arts, Braunschweig, Salon, p. 207. (translation by M.-E. Wachs, 2020).

2.1.3 Challenges: German and international provocations in Industrial Design

In summary, the transition from practising as industrial designer ('Industrieller Formgestalter') in the past, to becoming design theorists in the present, reveals an obvious effort to develop more and more into 'design scientists'. But which historical insights has this research on Industrial Design at the Braunschweig University of Art revealed? What is the goal, now that research has been done? The physical archive has been examined, interviews have been held with professors and alumni, statements have been given by alumni from different study generations, and best-practise designs have been documented by means of 'images'.
In addition, the author was able to draw on her experiences as a former student (1995-2001), and her experience as a former candidate of the doctoral programme in the interdisciplinary fields of industrial design and cultural sciences (2003-2007). She also gained insights into other regional universities in Germany and abroad in Great Britain and Sweden. All this knowledge was collected to provide advice to the universities and political parties of our European societies.

Germany faces the same challenges in industrial design or design engineering as European and Western countries (the difference between the names of study programmes depends on the defined profile of each University). Much of the following advice offered by Wachs can be transferred to varying degrees to other Design Engineering places of study in Germany and abroad. This will become clear as you learn more details in the upcoming chapters – particularly in the 'position paper' in chapter 7.

After the review in chapter 2.1 and 2.2. and the parallel research that took place, the following are the predominant challenges for the education systems in industrial design and design engineering, on which we should focus in the coming years:

— Designing trans- and interdisciplinary heritage – for example, by using the cultural heritage found in collections of museum objects, that are incorporated into art and design, in communication design and industrial design campaigns, for museums, and other representative design places of the 21st century

— Industrial Design as a brand – through the use of branding in the broadest possible sense in respecting 'art' as essential trigger and educational point of view

- Promoting competencies – such as by also highlighting the postdoctoral qualification in design as a unique feature of the university's profile. In Germany, there are not enough qualitative postdoctoral institutions that provide a good foundation for postdoctoral work as well as for 'mentee programmes'
- Respect gender, establish a programme for innovators and drivers; meaning social support for the equality of socially disadvantaged
- Universal design spaces for the practical use of artefacts: refocus regional strengths and re-evaluate the profile, offer a wide variety of workshops: from classic model building with clay, wood, metal, glass, porcelain and the use of CAD, through smart materials, all the way to the Augmented Reality studio
- 'Thinking workshops' that are international and interdisciplinary in their nature and cover topics from art, design and the sciences
- Study programmes in English and other languages as a new normal
- Encourage and support design and design education, in the form of a design foundation, which maintains and cultivates (the Braunschweig University of Art's) cultural heritage, as well as creating regional and national profiles. These should be intensively promoted and made available throughout Europe, and design skills should be conveyed much earlier in schools, thereby promoting accessible cultural education
- More in-depth regional research, as well as in Lower Saxony, communication throughout Europe and internationally
- Transdisciplinary sciences should collaborate with nearby universities supported by digitally interlinked collaborations throughout Europe and abroad
- Increased cross-generational teaching and development.

2.2 A 'contract' for cultural education – the political willingness

(Art)colleges have a responsibility, in fact a 'contract', to educate – now more than ever! And, not just with respect to the vocational qualifications of future artists, designers, and scientists – including all the soft skills that they need to develop on a personal level. They are also responsible for providing specific cultural educational opportunities that are part of a global 'studium generale' all the way to a professional degree, and frames and substantiates a unique qualification within a regional, European and global network.

In the first volume of the commemorative publication '50 Jahre HBK Braunschweig – Geschichte einer Kunsthochschule und ihrer Vorgängereinrichtungen', Claudia bei der Wieden outlines the efforts of self-reflection within the community of colleges – efforts which manifested themselves at the beginning of the 21st century. (In this process, she pointed to an exceptional group of alumni designers and their careers as professors at universities during that time.) She went on to say that *'Aside from socio-cultural factors, this is not least due to legal and political aspects: In consideration of their specifics, art colleges always represent a segment of the general education policy.'*[35] Cultural heritage, which is generated through design, impacts society and simultaneously reflects that same society. Thus, we should prepare, maintain and share a broad-based analogue and digital 'Knowledge Bank' (Wachs, M.-E.). This serves to secure our material and immaterial cultural memory – our archives – that should be made available for many future, diverse generations to enjoy. It will encourage them to ask questions and help them find answer regarding the 'shape' of tomorrow's culture – that means also questioning 'form' in a metaphorical sense.

Let us think of these collective thoughts, that derive from a collection of 'similar images', as a mind map – **a chance for stronger cultural education within Europe, as well as improved competitiveness of European 'engineered art', combined with 'design attitude' and the skilful 'creative ability' that the discipline design reflects**. This can be supported through learning and growing by means of generating your own cultural heritage – your own cultural character as compared to others and welcoming 'otherness'. We should continue the work on design heritage, which has just begun, and expand the interdisciplinary (art) collections within the region. In addition, alumni collections, not just those alumni working in international design spheres, should be maintained to create a broad-reaching institutional profile. The task is to make the characteristics of contemporary design visible – with the help of cultural memory, among other things –

[35] See Bei der Wieden, C., 2013, in: von Amelunxen, H. + Bei der Wieden, C., HBK BS (Hrsg.), 2013, 50 Jahre HBK Braunschweig – Geschichte einer Kunsthochschule und ihrer Vorgängereinrichtung, p. 227, translated by L. Huber-Fennell, 2021.

and to design the future: 'industrial' design products, concepts, science and design culture, probably with a view on political support.

Looking to the industrial design (education) heritage – in history, as well as the present time – it helps us explore the questions of 'form', in order for us to have a 'grasp' of both material and immaterial design characteristics and qualities used in the creative realm of experts. This is how future design codes will be generated, which can be used to represent valuable design heritage at a later point in time. This will serve as a foundation for future cultures to 'read' and build upon.

As such, they serve as signs that we incorporate into the design of a product, a room, a communication concept, or graphics, as well as in workshops, architecture – even spaces in a metaphorical sense. *'The spirit of art is created by hand' (Lütgens, A., 2012)* [36], and the same applies to design. In fact, it is precisely these design codes and media, that have been used since the 1960s to analyse and shape design theory: 'a system of signs' and 'semantic meanings' (see Barthes, R.). 'Design insignia' serve as representatives and showing us the design worthiness.

With regard to the question at the beginning of chapter 2, the citation by Karl Wollermann, from 1953, may well ring true again in 2022: *'At this school, the handcrafted work is to be given the nobility of a genuine form culture, which must first assert itself in the Brunswick and Lower Saxony areas, in such a way that together with other art schools form values are created, which later transform into economic values'.* The profile demanded of our society's – and our planet's – next 'complex problem solver' tells us that designers and design engineers need to maintain an open minded, holistic view and possess, a broad range of skills, to produce 'good forms'. This refers not only to problem-solving abilities related to product silhouettes and inner/outer qualities, but also the designer's consciousness and ability to give meaning to design solutions for the problems of the world.

---

[36] See Lütgens, Annelie, 2012, Fontanas Schnitte. Vom Leinwandbild zum Minikleid, in: Geiger, A. and Glasmeier, M. (Hrsg.), 2012, Kunst und Design. Eine Affäre, textem, p. 106, translated by M.-E. Wachs, 2021.

IMAGES 01

Industrial Designer of HBK Braunschweig: Design success today

In the year 2013, as former industrial-design student and PhD candidate of Hochschule für Bildende Künste Braunschweig, I was invited by Hannes Böhringer and Michael Schwarz, to look at the archive of the HBK Braunschweig – within the history of industrial design education to formulate a position. Recognizing so many fames designer from this fine place of creativity, it was my choice here, in this post doc thesis, to give an impulse for all universities of design disciplines and fine art: to take this beneficial cultural 'good' – our design heritage in form of legacy data in the archives of universities – more serious, to care about and mediate to the present and next generation. It is a contribution to 'knowledge building management' research in Europe', which is my interest to support by participating at the EPDE Conferences – representing Germany's Design Engineering community over the last years.

With the help of alumni contributions, you will find a small selection in the following pages, you could follow the development and I would like to invite you to enjoy, being curious about the development in design engineering product and processes, when some of the alumni of HBK Braunschweig will give the opportunity in documenting the design process, media and products as state of art in the final year of their studies compared to creative fields in the present time – enjoy a living testimony and futures design heritage… Thank you so much to all the willingness of the designer to let us participate and to give an insight. The following documentation relates to the chronological order of the final year of study of the alumni.

Wolf Karnagel
Klaus Zyciora (*Bischoff)
Kora Kimpel
Florian Altmann
Markus Rudlof
Markus Schweitzer
Marina-Elena Wachs
Andre Franco Luis
Alice Kaiserswerth
Manuel Windmann
Stefanie Krücke
Nicole Losos

Wolf Karnagel (*1940)

Diploma 1962

Studied at Prof. Bodo Kampmann, 1962, at HBK Braunschweig. Find some picture by his design in forming porcelain and metal – a precise form language and – heritage of German design. Wolf Karnagel lives and works in Berlin.

More 'Stambul'-porcelain series you find at KPM – Köngliche Porzellan Manufaktur Berlin.

A

B

A
Wolf Karnagel Design, 'Mokka Service Stambul', StPM, 1967.
Picture: Ulrich Fischer / Stiftung Leuchtenburg

B
Wolf Karnagel, porcelain dishes and silverware for Lufthansa – first class, 1985.
Picture: Ulrich Fischer / Stiftung Leuchtenburg

We thank Foundation Leuchtenburg, Seitenroda – Germany, for making the data (pictures) available: thank to Dr. Ulrike Kaiser, director of Foundation Leuchtenburg.
www.leuchtenburg.de

Klaus Zyciora
(geb. Bischoff)

Head of Volkswagen
Group Design

Contact
klaus.zyciora@volkswagen.de

Diploma 1989
Corrado Interieur Design

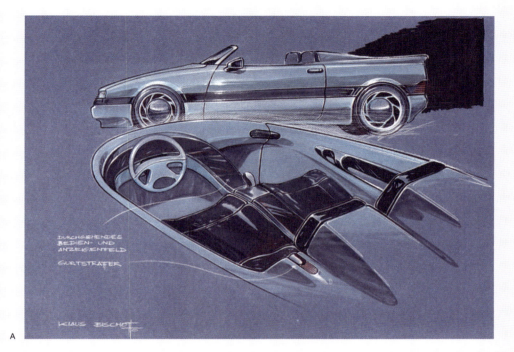

A
Diploma Corrado Interieur Design.

B
ID Buggy, 2020, VW represented by Klaus Zyciora, Head of Volkswagen Group Design.

Kora Kimpel

**Diploma 1995
Design of a portable school laptop with double screen**

Professor for new media at the Berlin University of the Arts

Kora Kimpel is a new media designer and has been teaching as a professor at the Berlin University of the Arts since 2004. Her research focuses on new forms of interfaces and interaction patterns. Another area of interest in her work is the development of methods and procedures in creative innovation processes. Kora Kimpel studied industrial design with a focus on interface design at the HBK Braunschweig.

Contact
mail@kkimpel.de
www.kkimpel.de
www.iid-udk.de

A — C
Diploma 1995
D — F
Recent Projects

A
Prototype in one-screen mode with CD-Rom.
Photo: Lutz Bertram

B
Prototype opened with 2 screens.
Photo: Lutz Bertram

C
Sketch of different usage situations of the laptop.

A

B

C

Kora Kimpel

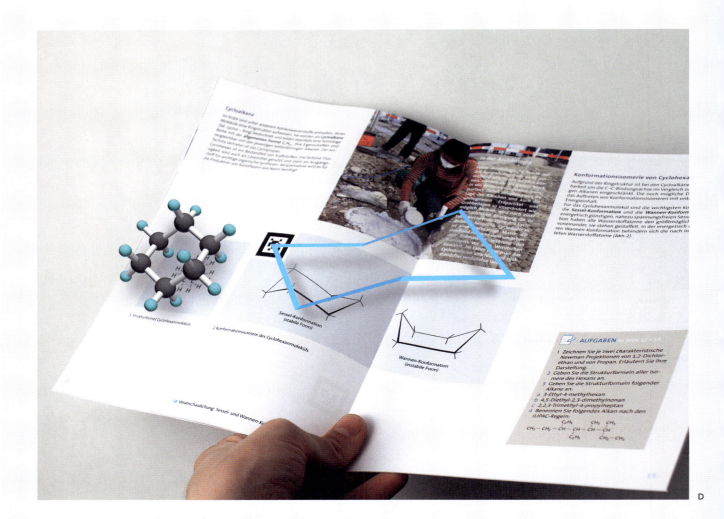

D
Chemistry textbook with augmented reality extensions for the Cornelsen publishing house.

E / F
Models for a shape-changing interface in the research project 'Smart Material Interface'.
Photo: Linda Elsner/ Joanna Dauner

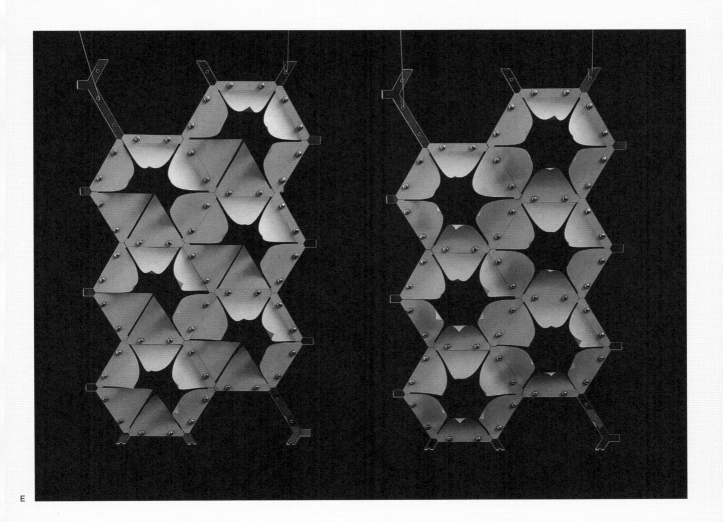

E

New interfaces will communicate directly through the material. Smart materials are the first step towards these 'material user interfaces'. The design of materials in terms of sustainability, interactivity and communication will be a new field for designers in the future.

F

Kora Kimpel

Florian Altmann

**Diploma 1996
SKID – Sportgerät für KIDs –
sports device for children**

Talent Group Lead
Customer Experience,
Applied Design and Innovation
at Deloitte Deutschland

Contact
faltmann11@gmail.com
https://www.linkedin.com/in/florian-altmann-025665/
Deloitte: https://www.linkedin.com/company/deloitte-deutschland/

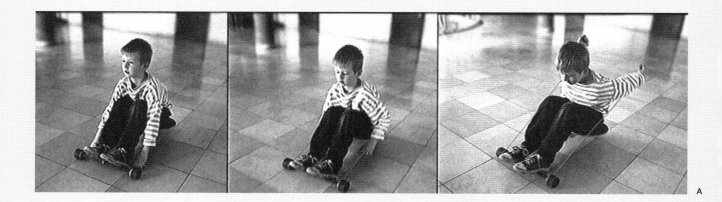

A

A — E
Diploma 1996: SKID,
Sportgerät für Kids – prototype.
All pictures: Florian Altmann

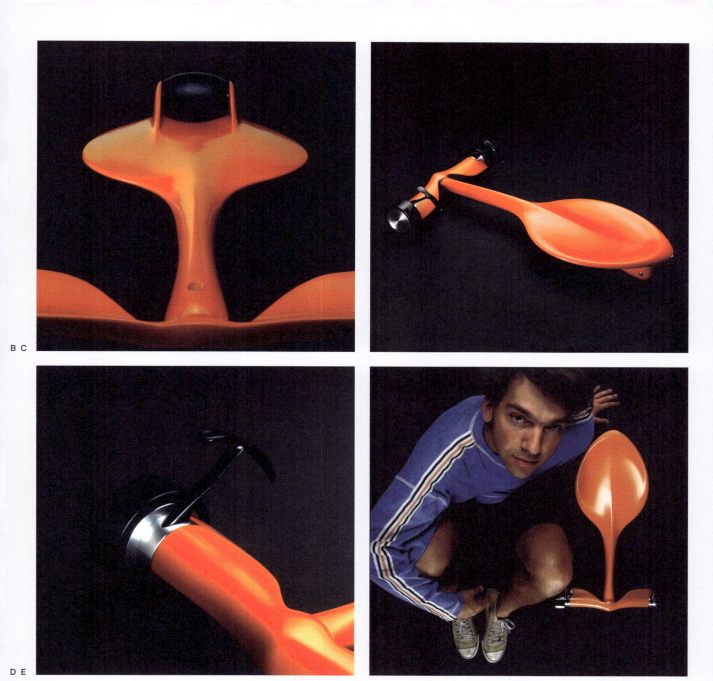

B C
D E

Florian Altmann

F G

F — H
'choreo – digital choreograph'
1995, as pre project to the diploma,
in developing the design concept
and producing the 3 d model.
The design is focussing on user
experience and form.
All pictures: Florian Altmann

H

Markus Rudlof

**Diploma 1999
Bentley Crossover
SUV/Limousine**

Markus Rudlof works as Industrial Designer within the automotiv industries today. The pictures and model is taken from the time of studying industrial design at HBK Braunschweig.

A

B

A
Bentley Crossover SUV/Limousine.
Pictures by Markus Rudlof

B
Monoposto Lightweight vehicle.
Pictures by Markus Rudlof

Markus Schweitzer

Senior User Interface Designer

Contact
schweitzer@faceinterface.de

Diploma 2001
Interface for monitoring complex process

The interface relies on a simple visual pattern to visualize complex data relating to automated industrial processes and render errors visible. Key messages need to be filtered out of the mass of processual date in order to give them priority treatment. By varying the basic structure of the visual pattern, the interface ensures different statements can be transported. Starting with Landolt rings, which are characterized by an opening in the ring, an individual message profile is produced. The 'event rings' can depict up to eight different parameters. The importance of each message can be symbolized either by varying the size or the strength of the lines.

Characteristics of the event rings:
— due to the visual pattern users perceive errors intuitively
— users can recognize affected process parameters directly
— users can compare the urgency of messages immediately
— users get an intuitive high level profile of the process condition.

Designing is a means to deal with uncertainty. By designing you take the conscious, and reasonable decision on one preferred out of multiple possible futures. This implies responsibility for society and the need, and openness to continuously rethink decisions, both small and large scale.

A
Exemplary assignment of 8 parameters to the different opening positions of the event ring.

B
Urgency of errors visualized by event rings over course of time.

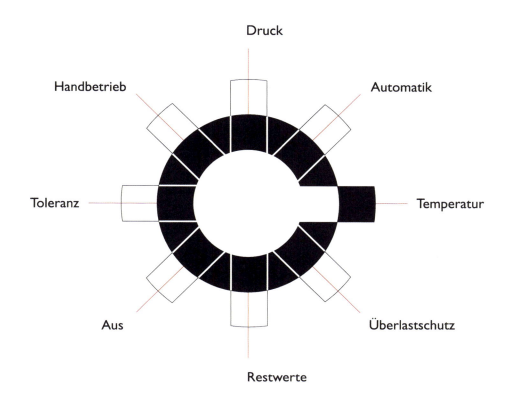

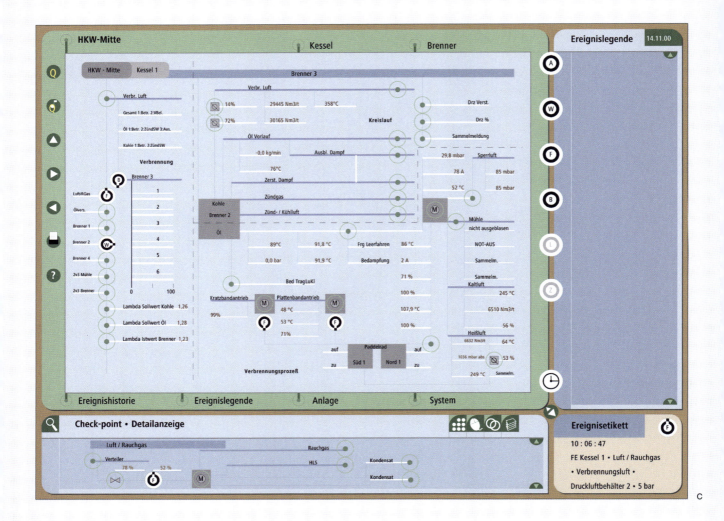

C
Exemplary UI visualization
for a control room.

D
Different visualization options –
optional white background,
prioritization by size or by
line offseta NC woodworking
machining center: Touch interface
for an embedded system,
in collaboration with Siemens
experts © Siemens AG 2021.

E
Exemplary assignment of
4 parameters to the different
opening positions of the
event ring.

Design and picture rights:
Markus Schweitzer

D

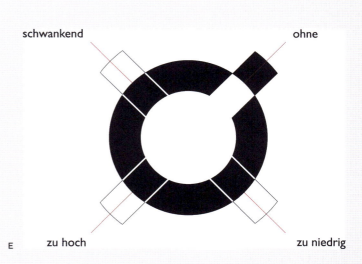

E zu hoch   zu niedrig

Markus Schweitzer

Marina-Elena Wachs

Marina-Elena Wachs Prof. Dr. phil., Dipl.-Des. is master tailor and industrial-designer and professor for design theory. 2007 she graduated at Braunschweig university of art with doctoral thesis ('material mind – new materials in design, art and architecture', https://verlagdrkovac.de/978-3-8300-3292-2.htm). Since 2010 she held a professorship in theory of design at Hochschule Niederrhein.
Marina's last research publication for EPDE Conference 2021: 'Self-confidence & self-expression through sketching – the significance of drawing in 'primary education' & the next generation of engineering'.

Contact
info@marinawachs.de
www.marinawachs.de

A
Time less design of a classical white coat: Marina-Elena designed and tailored this master piece of haute couture in the year 1986, Hannover, constructed by Marina-Elena Zieger (Wachs), photo: 2021, Velle. She is wearing it until today – a statement of time less and sustainable design.

Diploma 2001
interior / EXTERIOR –
the poetry of surface

The diploma by Marina-Elena, with the title 'interior / EXTERIOR – the poetry of surface' represents the break-up of disciplinary conventions on 'designing', underlined by empirical studies of the interior designer Patricia Guild, the artist Francis Bacon and the architect Richard Meier ('designer of interior and exterior'). A design theoretical approach at the one side, in using a holistically view on creating good, and a philosophical exercise at the other side, shapes an application on designing the usage of a 'hand-bag' in three different materials and form variations. Because the inner side of the individual is creating by hand and offering the inside out, touching is an emotional act and grips into the design process; the product language could be transferred to other designing languages. The diploma from Marina-Elena Wachs was coached by Prof. Dipl. Industrial-Designer Erich Kruse and expert Dipl. Industrial-Designer Ulrike Brandi; Univ. Prof. Dr. phil. habil. Hannes Böhringer consulted for the philosophical expertise and of Art Science.

Interdisciplinary design by 'light goes!'- design group 2009

Lighting planner Ulrike Brandi, architect Ulrich Tillmanns, architect and industrial designer Manuel Windmann, industrial designer Marina-Elena Wachs, founded together with another person the cross disciplinary design studio Light goes! In an experimental room (lab, garage;)) we designed lamps made of wood, glass, translucent concrete, with the demand to real good light. Thank's to all of our group for this fantastic experiences and learning 'space'. Thank's to Fritz Becker GmbH & Co Kg at Brakel, Germany and to Joachim Schelper and his always open mind and willingness to create long lasting designs underlined by wood and textile.

Time less design 1986 until today

Time less design, created by hand and mind: underlined with this case study of the classical white coat, sustainability is part of my live with consciousness and latest to express in designing good in the year 1986 (before I studied industrial design from the end of the 1990s on).
This off-white blazer coat – tailored of wool, inlays by natural materials (e.g. woven fabric of horse hairs – as well the classical tradition to hold the form of the revers) and horn buttons – it is designed and hand crafted by Marina-Elena Zieger (today: M.-E. Wachs) during the time of second apprenticeship year for learning the craft of haute couture at Atelier Behrens, Hannover, Germany.
A white trouser in Marlene Dietrich style complemented at the same time, made of the same peace of woolen fibers of first quality. Today it is an important testimony of textile heritage made to measure for the human body, but at the same time a certificate of our 'knowledge management' and 'building knowledge' to mediate to the next generation, in a resilient manner. Over the last decades! this classical white coat was and is my favourite of my clothes – my second skin that I am wearing, which I am following with passion and care about very well.

A

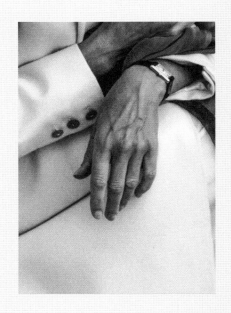

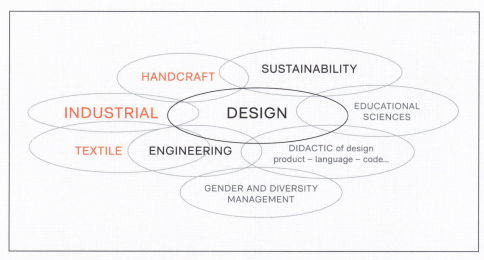

B
Maternity area designed by Marina-Elena Wachs, a study during her industrial design study programme, pre-diploma, 1996, photo: M.-E. Wachs.

Graphic 02
Business and research fields /- development by Marina-Elena Wachs, M.-E. Wachs, 2022.

56   IMAGES 01

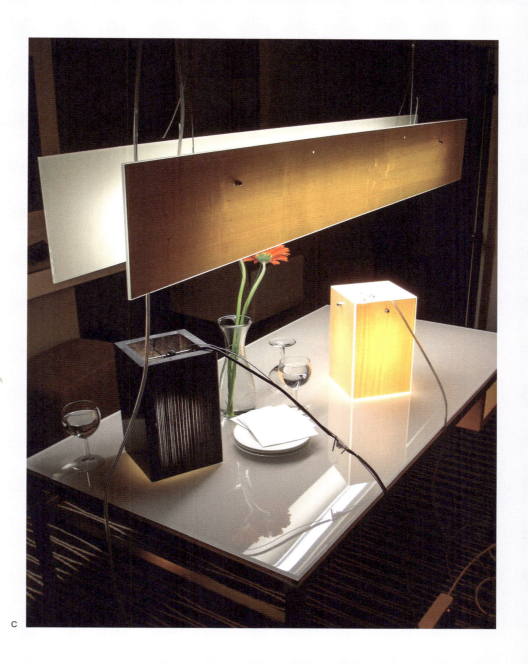

c

c
Lamps designed by the inter-
disciplinary design group 'light goes!'
By: Ulrike Brandi, Bernhard Mann,
Ulle Tillmanns, Marina-Elena Wachs,
Manuel Windmann.
You see the lamps: 'Light goes on
wood', 'Light goes on concrete',
'Light goes on wood and glas',
during a presentation 2009, Berlin.

Thank to:
https://www.ulrike-brandi.de/en/
https://4a-architekten.de
(Ulle Tillmanns)
https://marinawachs.de/en/intro-e/
info@manuelwindmann.de

Andre Franco Luis

Former Director,
Interior & UX/UI Design
Karma until March 2020,

Senior Interior Design Manager at Arraival since April 2021.
Irvine, Carlifornia, United States of America

Contact
andrefrancoluis@icloud.com

Diploma 2005
Interactive networked automobile

INA Diploma thesis / The diploma thesis Interactive Networked Automobile (INA) in collaboration with VW DESIGN represents the early ambition to create a vehicle Interior that dissolves the conventional status quo of vehicle architecture. It also questions the need for further connection and network in between the vehicle and it's environment while including the mobile phone as a central hub.

A — B
Diploma 2005
C — D
Recent Project

A / B
Diploma interactive
networked automobile.

A

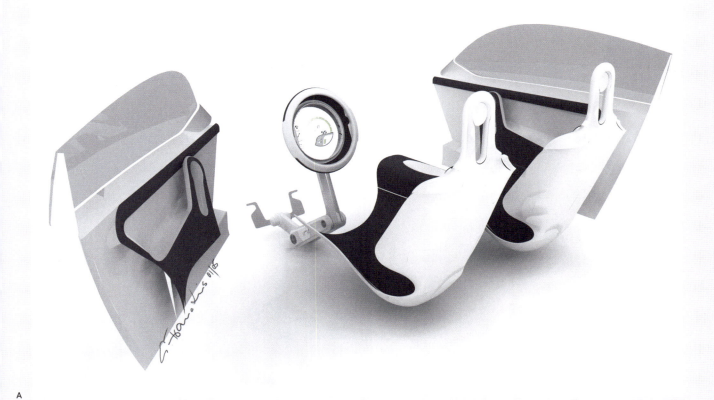

B

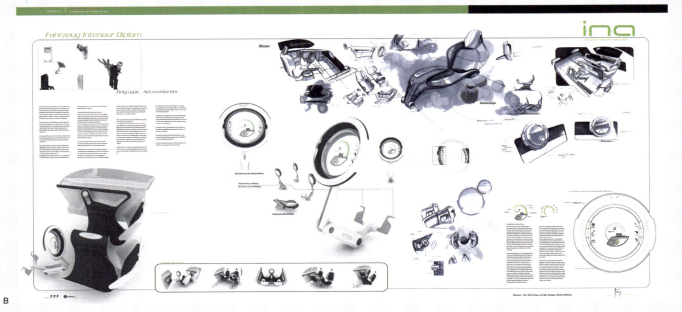

Andre Franco Luis

D

C
Karma SC2 interior sketch published by the show car launch, 2019.

D
Karma SC2, 2019.

Picture rights: (A – D) Andre Franco Luis and in cooperation to each enterprise he worked for.

Alice Kaiserswerth

Senior Footwear Designer, Reebok Intl. Ltd., Boston, MA, United States (since 2008; since 2012 home office from Munich, Germany)

Contact
alice.kaiserswerth@gmail.com

**Diploma 2006
Urban Croquet –
a new way of lifestyle**

Urban Croquet is a concept, developed in collaboration with Adidas Sport Heritage. Freed from the groomed lawn and the strict rules of the traditional sport this reinterpretation of croquet becomes a new outdoor experience and an expression of lifestyle.

The entire collection consists of a game set, apparel and a shoe. All elements of the game are designed to be used on the multiple surfaces that exist in urban surroundings – like parks, parking decks or empty streets. A clean design language using iconic colors and recognizable shapes unites all elements into one cohesive collection.

A — C
Diploma 2006
D — E
Recent Project

A
Game set consisting of mallet, hoop and ball (3D model).

B
The game pack – 4 mallets, 4 balls, ten hoops and 2 sticks in a bag (3D rendering).

C
Ideation for the design of the mallet.

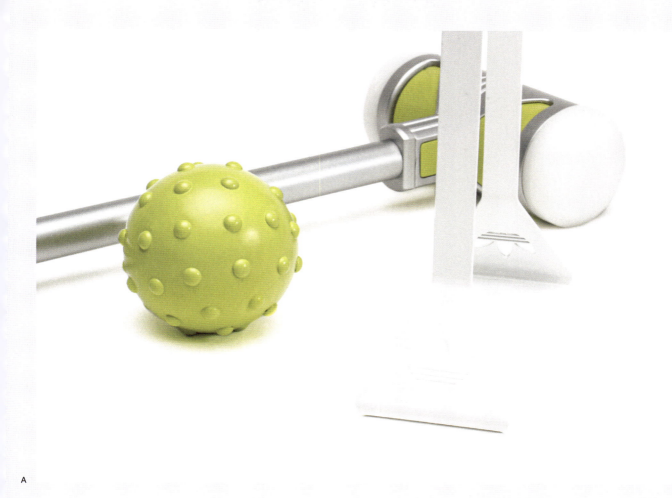

A

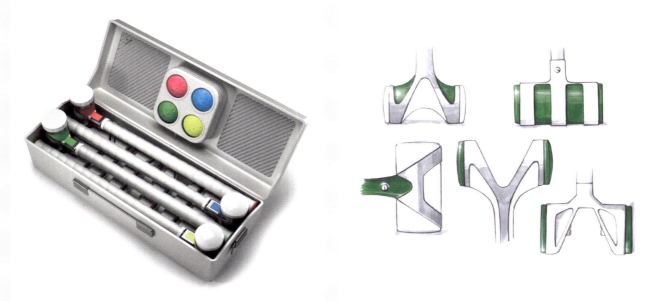

B C

Alice Kaiserswerth

E

D
Reebok Z-Series Run Season
SS 14, category running.

E
Reebok Esoterra: season
FW 2017, category walking.

A – C: all pictures by
Alice Kaiserswerth
D – E: Alice Kaiserswerth /
Reebok

Alice Kaiserswerth

Manuel Windmann

Architect (graduate engineer)
Urban planner (graduate engineer)
Industrial designer (graduate designer)

Contact
info@manuelwindmann.de
www.manuelwindmann.de

Diploma 2006
Le restaurant d`hôtes (2006)

Le restaurant d`hôtes as diploma (2006) of architect and industrial designer Manuel Windmann, is documented here with a still of 3D animation construction (pic. A) and a hand formed prototype of a detail (pic. B) of the concept. Designing interlinked, with different laboratories at HBK Braunschweig was a great benefit during the study time.

The pharmacy at the Airport Hannover (2009) is presented to customers as an accessible sculpture. A spatial element of the pharmacy is the long, curving sales counter, which is a main presentation feature of the space, nestling up to the rounded walls of the laboratory and consultations areas. This expansive element is made from a three dimensional solid surface and divided by a horizontal orange stripe, thus developing a coherent character. The colors and surfaces design of the pharmacy are deliberatly minimal, allowing the wares to be optimally displayed. The ceiling lighting gives the impression of daylight and creates a brightly lit atmosphere that supports the flowing lines of the design.

A — B
Diploma 2006
C — E
Recent Project

A
Le restaurant d`hôtes (2006)
Final design, industrial design concept, design, 3D-modell and visualizing.

B
Design study (2006)
Plaster model and sketches.
A – B picture rights:
Manuel Windmann

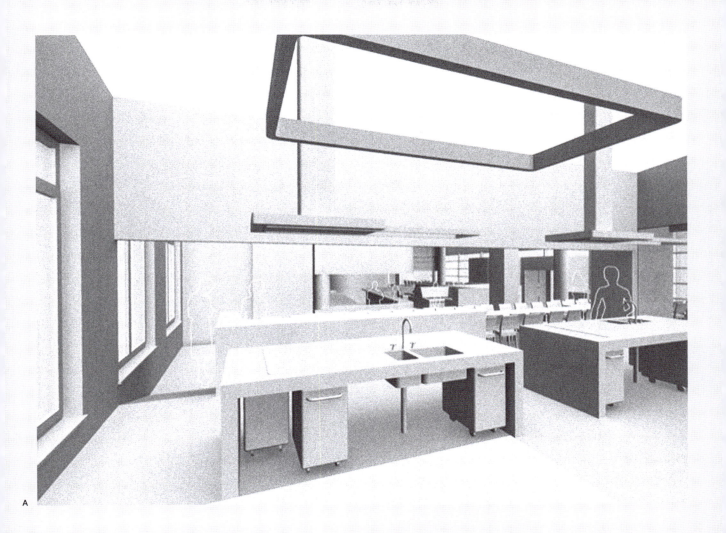

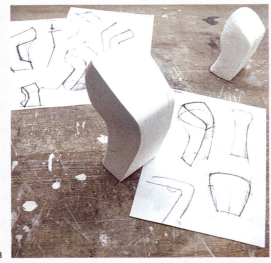

Manuel Windmann

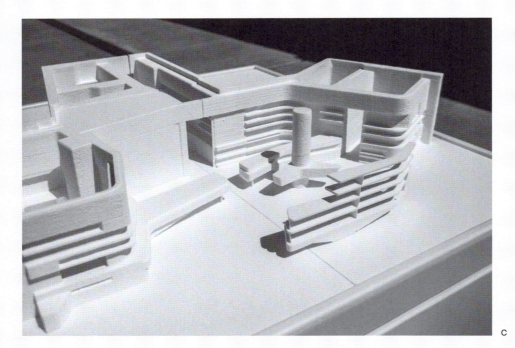

C
Airport Pharmacy –
Airport Hannover (2009)
Plaster model.
Picture rights: Florian Holik
und Manuel Windmann

D
Airport Pharmacy – Airport
Hannover (2009). Sales
counter with structural orange
elements and handbag-rest.
Picture rights:
Hans-Jürgen Oertelt

E
Airport Pharmacy – Airport
Hannover (2009). Sculptural
solid design-elements,
sculptural adjustable shelf,
the ceiling lights give the
impression of daylight in
the middle of the airport.
Picture rights:
Hans-Jürgen Oertelt, Hannover

Stefanie Krücke

**Diploma 2007
KITANO Kite Sailing Yacht**

Dipl. Industrial Designer
Since 2013 Head of Design
Bavaria Yachtbau GmbH,
Giebelstadt

Winner: 2007 Concept Boat
Award - London Boat Show
London, GB

'Kitano Kite Sailing Yacht' -
Overall winner + Pure Concept
category winner

Stefanie Krücke

Nicole Losos

Contact
Nicole Losos
http://www.losos.de
info@greenhousesystem.de

Werner Aisslinger
studio@aisslinger.de

Diploma 2009
BergFrei as concept design

The diploma BergFrei concept (2009) was designed at HBK Braunschweig in regard with coaches of the University.

The design Yill is developed in cooperation with Werner Aisslinger at studio aisslinger, Berlin. Yill is created as energy storage system for the enterprises Younicos AG.

A — C
Diploma 2009
D — F
Recent Project

A
Screen topographicouch BergFrei.

B / C
BergFrei, Diplom, Rendering.
A — C design and picture rights by Nicole Losos

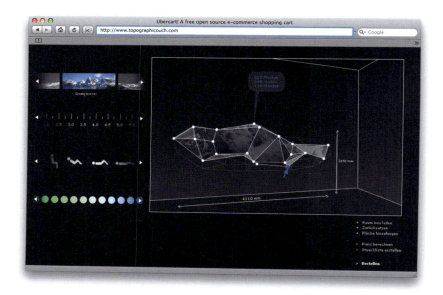

A

B

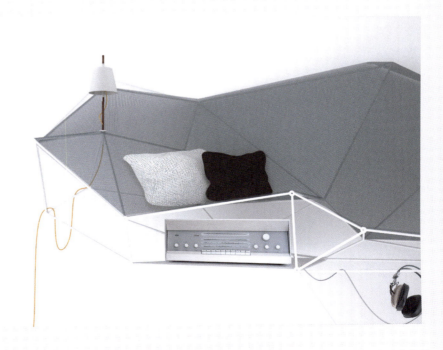

C

Nicole Losos

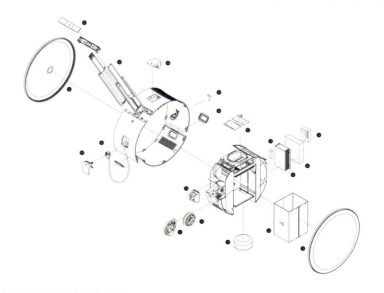

D / F
Cooperation: Werner Aisslinger
& Nicole Losos, Berlin,
Yill Produktfotos – CREDIT
YOUNICOS.
Picture rights: Werner Aisslinger,
Nicole Losos, product picture,
credit YOUNICOS

E
Exploration sketch Nicole Losos.

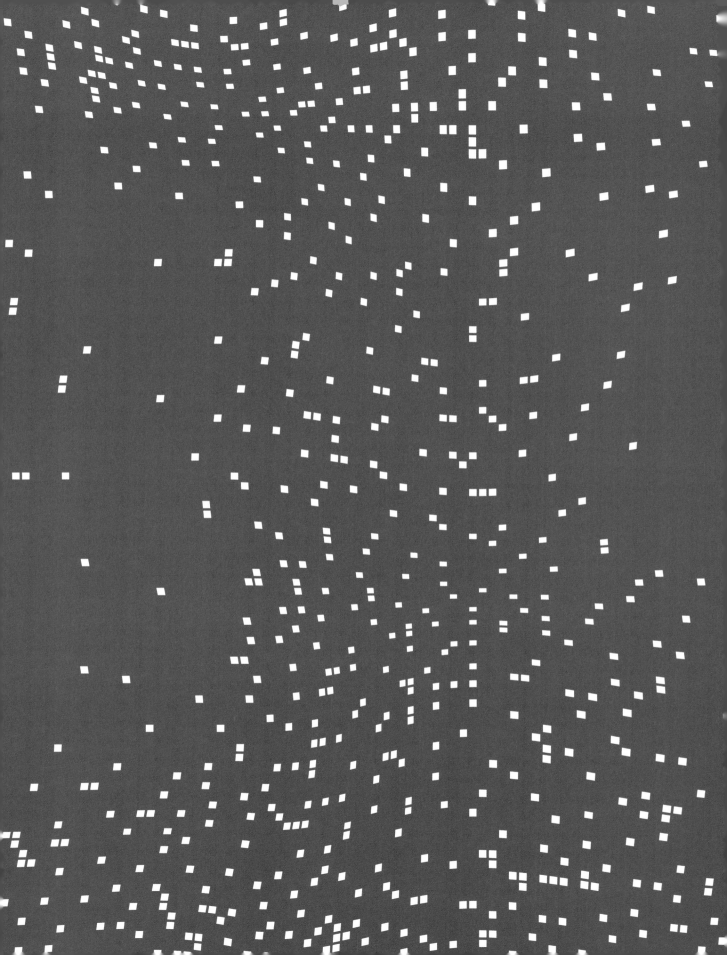

CHAPTER 03

76 — 101
The design term 'textile engineering',
the significance of sustainable
design drivers and the 'design shift'

3.1 The driving forces of (textile) culture: 'Textile engineering' in history and German industrial design – a transdisciplinary look at working conditions, industrial changes and people's source of identification

3.1.1 Introduction: The case study 'Textile Traces in Germany' in relation to sustainable design studies

Keywords

textile design engineering, smart textiles – textile performance, tactile driven product language, textile engineers as drivers for sustainable design engineering solutions, textile nerds and textile wo/manpower, textile traces – individual path ways – places to remember – knowledge archives, textile heritage, textile sketching in the AR, to invest in human capital, osmotic engineering creative spaces, theory of creation and textile theory.

The previous chapters focussed on the characteristics of design education history, based on industrial design education AND the product language heritage of the German Werkkunstschulen, as they relate to industry. The combination of handmade functional 'hard' objects in industrial design and design engineering have shaped the foundation of this heritage. However, the next chapter concentrates particularly on *textile* design engineering. In the last decades the interest in smart textiles and 'textile engineering solutions' (Wachs, M.-E.) has expanded, because textiles as a material have become increasingly significant for future multi-perspective sustainable solutions. This was the case then, especially those relating to tactile-driven product language – and the female and male drivers in that case – and has continued till now.

The case study 'Textile Traces in Germany' is part of a larger research project 'sustainable design studies – within the scope of cultural theory and material behaviour', by Marina-Elena Wachs. This study incorporates multiple perspectives, beginning with industrial design, and continuing through handcrafts and manufacturing, to the specialization of design theory used as a method in design engineering in Germany. Subsequently, this was compared to Europe studies. In addition, the cultural anthropological views that emerged as specific regional expertise came into focus: In this case, it was the expertise found in textile industries and engineering history in Krefeld and Mönchengladbach (However M.-E. Wachs does not end in generating new places and 'spaces' as learning and working landscapes, with the help of a tool for 'textile sketching in the Augmented Reality' in 2021).

The research study, 'Textile Traces in Germany', is focussed on education for sustainability and the changing working conditions in the textile industry. This is related to how the workers who served as driving forces identified with their region in former times. **The study also compares Germany to textile centres and the great textile expertise found in other European countries. The long-term research project is particularly interested in intercultural codes marked by the typical characteristics of creativity and the creative process.** These codes result in insignia – the representative icons of each society. This heritage represents the 'cultural memory' (Assmann, J.), and is based on the individual paths people took in life, and the impact of (industrial) developments.

These typical paths in life are simultaneously evidence for textile experts' passion, rooted in handcrafted and industrial processes developed over the course of history. Yet, they are also the basis for the cultural competence implemented in the textile industry's circular economy – both then, now, and in the future. We must look to the past in order to move forward in sustainable design areas. It is obvious that the textile industry empires of the past – for example Manchester in Great Britain, Borås in Sweden, Krefeld in Germany, St. Gallen in Switzerland and Lyon in France – were centres for manufacturing and industry. They had tremendous economic power thanks to the individual experts' skills and 'textile nerds' that created the textile manpower. The great passion for changing product language styles is facilitated by the beginning of a new consciousness of design. Similarly, the Bauhaus women in Dessau and Weimar created things together with the male pioneers of that era – their teachers – which generated a new style after the World War II.

In consequence, an examination of the relationship between various aspects revealed the interconnectedness of industrial development and changing social conditions: The relationship between the industrial heritage of steel and textiles (see, for example, 'history of technology', Heßler, M., 2012, Kulturgeschichte der Technik), the relationship between men and women, changing role models in business and working conditions (see Funken, Ch., 2016, Sheconomy – Warum die Zukunft der Arbeitswelt weiblich ist), the relationship between people and things (see Schiffer, M. B., 2009, People and Things), and, lastly, regional and cultural identification conditions (see Nowotny, H., 2005, Unersättliche Neugier – Innovationen in einer fragilen Zukunft). At the same time, the need for education relating to sustainability issues became evident.

Ergo, the study team always has to visit so-called 'lieux de mémoire' (Nora, P., 2005 (1997), Les lieux de mémoire) – places of memory, like Pierre Nora originally described with regards to France. It usually begins with an excursion to the Museum Zeche Zoll Verein in Essen for the students of the textile and clothing technology department, who are taking the design theory and history course. The museum is a place to remember – a place where the industrial and textile industrial revolution is made tangible. This place is characterised by the typical brick architecture of the industrialisation period, and is marked by machines used to produce coking coal for the iron and steel industry. This was the prerequisite that enabled the textile industry to evolve. Today, the Red Dot Design Museum is located at the Zeche Zoll Verein in Essen, Germany. Here the annual awards for good and innovative design solutions from every design discipline are on display. And you can discover a great collection of the German design and product history – in addition to our cultural heritage. This is one example of a place to remember for the German industry's history, as well as the first industrial revolution.

This retrospection helps, as we are now facing the 'Fourth Industrial Revolution', which brings with it, new challenges for the so-called Industry 4.0. Beside accelerating towards an enormous replacement of workers and tasks in steel production, related to the covid-19 pandemic (see Lagarde, C., 2020)[37], the great paradigm of the current digitalisation revolution demonstrates that we have, once again, reached a great mile stone for new economic working conditions – also within the textile industry. *'In tomorrow's world, many new positions and professions will emerge, driven not only by the fourth industrial revolution, but also by nontechnological factors such as demographic pressures, geopolitical shifts and new social and cultural norms. Today, we cannot foresee exactly what these will be but I am convinced that talent, more than capital, will represent the critical production factor. For this reason, scarcity of a skilled workforce rather than the availability of capital is more likely to be the crippling limit to innovation, competitiveness and growth.'* (Schwab, K., 2016, The Fourth Industrial Revolution, p. 44 f) As Klaus Schwab's thesis suggests, and in light of the development and consequences resulting from the technological impact of the digital revolution, we have to focus on talents' capabilities, and thus in 'human capital' more than anything else.

---

[37] See Christine Lagarde, 2020, in: Siedenbiedel, C., 2020, „Die Transformation ist gigantisch", FAZ, No. 272, p. 29. (Frankfurter Allgemeine Zeitung).

3.1.2 Challenges for textile experts and the textile industry's responsibility in the future – parameters for sustainable economic & social conditions (not only in the textile industry)

The arguments outlined in the previous chapter underline the thesis that the greatest challenge faced by the Fourth Industrial Revolution, in Industry 4.0, is the demand for new skills and new European education programmes. It is also evident that these both need to be developed in a variety of fields – not only within the textile industry and learning spaces (please compare to chapter 4).

At the same time, we must look at societal changes and shifting individual needs, while aiming to provide education in a sustainable manner for textile and fashion design engineering, as well as management.

Therefore, since 2018, the research for the case study has been focussed on the following fields of interests: Firstly, it has been concerned with examining textile heritage, as well as the objects and working conditions of the past (e.g. with the help of the Bauhaus Archive, Berlin / Museumsinsel Humboldtforum and others abroad like in Manchester). Also, expert interviews, comparative literature and digital references from various archives and libraries in Europe and around the world have been incorporated. Lastly, information has been gathered during field studies performed at special places of heritage within the textile industrial, as well as modern-day locations in Germany's lower Rhine region, Manchester in Great Britain, Borås in Sweden, and St. Gallen in Switzerland. Respectively, the following parameters have been defined:

Parameters for sustainable economic and social conditions (not only for textiles)

— history / heritage of products, technological possibilities
— knowledge of how to create value
— differentiation of generation characteristics, skills and behaviour
— impact of different media – childhood imprint and cultural background
— geopolitical background – conventional paths in life
— chances for change are found in innovation – psychological impact
— (in)security in life, and migration
— trusting or distrusting technology and digital working tools of the future.

The focal point of this case study's research in textile engineering is to look at the macro-perspective's parameters, which is related to sustainable industrial design engineering fields. It examines the following:

- changes in technological processes and the effect on patterns created by the paths people took in life, and the question of how people (and talents) identify with their regional heritage
- comparable textile design studies within the scope of historical, sociological and anthropological evaluation
- the innovative use and development of materials and technologies for creating objects (and design), which has served as a driving force in the development of design, materials and technological innovations, and has also influenced material behaviour
- the correlation between design and engineering
- the correlation between material and behaviour in relation to objects
- sustainable education and working conditions as a necessity, with regard to change (and loss) of manual skills, as well as the need for incorporating analogue AND digitally-based education solutions
- osmotic textile (design) engineering spaces connecting different disciplines, generations and textile knowledge about the textile heritage.

The study 'Textile Traces in Germany' also focusses on and is supported by the literature. Here, disciplines like sociology, design history, cultural history of technology, cultural studies (anthropology) and philosophy are taken into consideration. This multifaceted perspective, underlines the interconnectedness of industry and societal conditions.

Europe's textile heritage is related to objects, things and the paths people were destined to take in life. On the one hand, the study traces the footsteps of the Bauhaus women and teachers, who enjoyed an enhanced status, while on the other hand it looks at the working conditions in the textile industry after the World War II: this perspective revealed the culmination of individual choices, fortune and fate.

Furthermore, these factors can be identified in other European creative fields, such as furniture design and interior design: When former female students and teachers of the Bauhaus emigrated to the United Kingdom, there is evidence that an osmotic exchange of contemporaneous styles, working conditions and design thoughts took place, within the European mindset. Comparable processes and methods used in design engineering and collaborative developments – resulting from the work and life conditions during and because of World War II – simultaneously influenced the results and learning methods (see Powers, A., 2019, ibid: Pepchinski, M. et al, 2017, Frau Architekt-Over 100 Years for Women in Architecture).

3.1.3 Sustainable design models and textile traces according to the heritage outlined in Schiffer's 'work and people'

By examining this special cultural textile design heritage, comparable European studies have shown that design is the mediator between products and their users, and is of interest to entrepreneurs and consumer groups. By using resource-efficient production circles, product value generates ethical content, in addition to creating functional and long-lasting cultural goods.

High-quality products are a consequence of sustainable designing, because it seeks out the origin, the needs and additive design qualities, while focussing on responsible behaviour. You have to think, produce and invest in a sustainable manner if you intend to implement responsible production and economic circles. This, in turn, requires developing and employing a circular thinking model. All of these findings are based on our textile and industrial heritage, as well as the paths that were taken in life by an enormous number of individuals and impassioned people (see Wachs, M.-E., 2008). The challenge for the next century will be to develop a connectivity (see Nowotny, H., 2005) among the collections and institutions in Germany that represent important elements of our 'cultural memory' (Assmann, J.). It would be equally valuable to link these to other national institutions in the textile creative industry, such as the Fashion Council Germany and Deutsche Gesellschaft für Designtheorie und -Forschung, and, for example, the Design History Society in Great Britain, the research institute of The Swedish School of Textiles – University of Borås in Sweden, and VIA University College at Herning in Denmark (Please note the guest statements by Anne Louise Bang, Textile-Designer from Via University College, Denmark and Charlotte Sjödell, Industrial Designer from Lund University Sweden, Wenche Lyche from Oslo Met (see IMAGES 03)).

As we move forward in Germany, it would make sense to fall in line with other European and international design centres which focus on textile heritage – especially in terms of educational models in textile branches. The present ongoing digitalisation is providing great opportunities and can be used as a good trigger point for facilitating this step.

Today, sustainable design engineering models are only be truly comprehended through a retrospective understanding of the significance of technological history. The First Industrial Revolution served as the driving force for the textile industry in Northern England until the middle of the 19th century, and encompassed all parameters: resources, technology, financial and human capital, transportation conditions, infrastructure, as well as business market conditions and the strategies employed. Many of these parameters converged, to create the conditions that facilitated the industrial revolution. This first took place in England, which is why Great Britain is called the 'First Industrial Nation' (Liedtke, R., 2012, Die Industrielle Revolution) and must be viewed in that context. Innovations were sought and developed in order to produce fine iron, and subsequently machines of steel and iron were built – innovations for which coking coal was a prerequisite. Thus, came the steam engine and more sophisticated functions of energy, which substituted man power and water power. This is illustrated by the flying shuttle, patented by John Kay in 1733, which sped up the weaving process and was a key factor for industrialisation (see Liedtke, R., 2012, Die Industrielle Revolution, UTB, p. 33). The investment in machinery in Great Britain was often weighed against the high salaries of manpower, yet ultimately both were essential for success. The power of financial capital first became evident when innovations changed the parameters. The Flying Shuttle, for example, enabled faster production with higher output. For a while, this had severe consequences on the life paths of individuals and society, **as many weavers lost their work after being substituted by machinery**. However, manpower regained significance later on – as human capital – after Samuel Crampton invented the Spinning Mule, in 1779. This led to a differentiation between weavers and specialists, who founded a so-called 'labour aristocracy' that produced the highest quality, fine cotton fibres for the textile industry.

    The textile industrialisation of Northern England and Manchester was unique, and maintained its stronghold for several decades, until their enormous technological knowledge resulted in a number of patents and subsequently spread to Europe. The great success of the textile industrialisation was based on the convergence of several innovations, the development of technological solutions, and the capital raised and used to invest in resources. Additionally, human manpower, modes of transportation, the infrastructure, factories and the creation of strong centres for textile trade played a significant role, as well in Germany – like it plays today again about the question of trade and transport in creating a 'new silk road project' since 2018 (the port of Duisburg, North-Rhine Westphalia Region came into focus again).

    Once transportation had expanded to become a global network – in terms of both resources and manpower – it served as the success factor for the First Industrial Revolution. Similarly, the rapidity of information 'transport' is currently facilitating the Fourth Industrial Revolution. The largest key factor for the present digital revolution is the amount of time required to evaluate digital data and integrate it into the right concept for the market. Today, it comes down to the quality of the data's 'value' and the criteria of the subsequent evaluation, which define the price of products and services. **Yet, how do we find the right parameters to evaluate the price of workers – or should we refer to them as 'thinkers'** – who serve as the textile industry experts of future markets? We have to ascertain the right media combination: When do we need analogue methods and when is it more efficient to use CAD systems, digital tools and the use of AI? What (social and cultural) impact will this have on **low-skilled workers and the highly-skilled and**

valuable textile experts if there are **only two groups of employees** in the future? *'This may give rise to a job market increasingly segregated into low-skill / low-pay and high-skill / high-pay segments, or as author and Silicon Valley software entrepreneur Martin Ford predicts, a hollowing out of the entire base of the job skills pyramid, leading in turn to growing inequality and an increase in social tensions.'* (Schwab, K., 2016, p. 45).

As we have seen in the past, the technological innovations of a particular period in time demanded special skills of human beings, such as when fine weaving systems created the need for a rich labour aristocracy developed from a simple workers group. This tells us that we have to look at the full spectrum of individuals' life paths to understand the chances they had and the challenges they faced. After all, each person grows up with a unique childhood imprint and cultural background, and sometimes (like today) this also includes a geopolitical migration flow. Each individual displays different cultural and material behaviour and interprets information differently, thereby acting in a unique manner according to their own feelings and consciousness. The fear generated by the First Industrial Revolution can be compared with the feelings of insecurity that we see today, caused by the vast spread of information on the world wide web, as well as the fear that robots will substitute human work skills. We are already observing the impact of home offices, for example, on individuals' life and career paths. But are local work spaces really no longer needed and can they be replaced by digital work options? The Fourth Industrial Revolution has the potential, to both promote trust in technologies – in robots' unlimited possibilities – **and** to revive analogue sustainable 'slow design' and engineering working conditions that foster a valuable and meaningful life on earth. Afterall, you need the collaborative spirit of engineers and designers, creating hands-on designs together – in a development process that is characterised by digital **and** analogous experiences. This is necessary to generate new textile engineering based materials for fashion, textile and automotive design for example, (see exemplary IMAGES 02, p. 104 ff). But will a digitally characterised working place (space) be an alternative, a 'place to remember', and how?

3.1.4 Comparable European studies examine textile themes related to people's heritage, design culture, and the question of identification

QR Code 01

Sustainable design models and textile traces – as described in the heritage of 'work and people' (Schiffer, M. B.) – were examined in some case studies within the scope of the project 'Textile AchiSculpture', in 2018 (see QR Code 01), and presented at the first conference of 'Textile and Place 2018', initiated by Manchester School of Art and Metropolitan University.

Comparable European studies have also looked at cultural textile heritage, and have identified it as being beneficial for promoting satisfying working conditions, which trigger innovation: Cultural textile heritage is driving the 'Textile Industry 4.0' in design, engineering and management.

It follows that theoretical concepts in design will drive the future of design. The interdisciplinary design project 'Textile ArchiSculpture' (2018/19) was connected to research work performed by M.-E. Wachs, highlighting the textile heritage in North Rhine-Westphalia and connecting the textile expertise of Europe. On the one hand, it analyses concise textile industry locations as places of memory, within the scope of the textile studies that comprise the textile heritage of technological history. On the other hand, it refers to the relevance of representative textile production facilities and their unique link to one another (please note the relation between textile, coal and steel industries, in the past). In addition, it reviews the textile engineering talents today – from their spirit and working methods, to their sustainable design solutions. This is directly translated into benefits for today's education, as well as the industry's state-of-the-art technological possibilities in 2021.

The characteristics of the textile heritage are visible in the cities' architecture, industrial culture, textile product design, but also in the resumés of dedicated textile sponsors and drivers for innovative design. All of these parameters relate to one another – and will continue to do so in the future digital (production) world. The topic 'Textile ArchiSculpture' relates to the textile industry's culture and textile architecture, and explores and draws connections between Europe's beacons in the textile industry: Krefeld, Bielefeld, Mönchengladbach, and Vordemrade in Germany, as well as St. Gallen in Switzerland, Borås in Sweden, Lyon in France, Manchester in Great Britain, and many more. These locations serve as research sites and continue to provide inspiration for new sustainable textile solutions. Subsequently, the task of bachelor and master students in design is to create sustainable (textile) design solutions, by transferring striking design elements into different applied forms (in this case study in a mobile textile architecture (for example, a pavilion) that showed a parallel to the development in fashion – in the form of sports couture).

Subsequently, these modern-day textile buildings were examined during the project's kick-off meeting, together with the managing director of the company Textilbau GmbH Hamburg, Kim Reinsch. Here they experimented with nylon socks – a classical, conventional method of architecture – in order to develop new shapes and new textile sculptures of architectural constructions.

Hence, playing around with textile sculptures today serves to incorporate textile industrial architecture research and thus the search for connections in North Rhein-Westphalia and beyond. Through this project we are connecting textile experts across generations and partners in the region, as well as within Europe (e.g. Great Britain and Sweden) for the digital textile (production) world of tomorrow, in the form of sustainable textile solutions. In this manner, design is connecting European textile engineering sites through their extensive textile cultural heritage, and Textile Archi-Sculptures are producing textile linkages and the (demanded) textile expertise for tomorrow.

In order to share these insights, these case studies were presented at several fairs between 2017 and 2019: in Stuttgart and Paris, as well as at exhibitions in Dusseldorf and Mönchengladbach, in Germany. Simultaneously, it was available via digital media (catalogues and websites, during competitions), for example on the department's website. We must relate to textile heritage and acknowledge the benefit this reflection will have on tomorrow. *'It is necessary to know the past to build the future.'* (Giorgio Armani, 2020) [38].

[38] Armani, Giorgio, 2020, in: Armani movie: Time less thoughts – the video – documentary, 2020, min, sec: 1,16, narrated by Pierfrancesco Favino.

3.1.5 Thinking ahead for the needs of society and the textile sector after the Fourth Industrial Revolution – especially sustainability

It is clear that the needs of society and the economic conditions after the Fourth Industrial Revolution indicate the parameters for the transformation process, as formulated here:

- industry transformation with consequences in production and services
- changes in consumer needs – less material based
- evaluation of sensible applications of digital AND analogue tools
- complex interdisciplinary and intercultural European tasks that should be solved together
- new 'togetherness' (Richard Sennett) – not only for designers and (textile) engineers
- geopolitical changes – new working conditions and educational challenges
- developing a positive attitude towards the fourth industrial revolution as paradigm change
- it is a chance for everybody to participate in creating sustainable living conditions.

These parameters of the change process relate directly to Marina-Elena Wachs's current research subjects:

- sustainable design / design management solutions
- fostering a unified Europe with the help of intercultural and interdisciplinary projects in design / design management – to reduce the increasing geopolitical problems
- underlining and supporting (inter)cultural education as the basis for design and design management
- focussing on challenges and chances of the transformation created by the digital revolution
- promoting an equilibrium in the gender gap, diversity management in design and design management
- enhancing new expertise within complex systems, which must be modified with new skills in a continual basis – beginning with children's education.

Due to the fact that **talents are our prospective human capital**, the European interdisciplinary and intercultural education programme is very worthwhile, and the best investment for Europe to obtain the pole position and steer clear of a geopolitical crisis.

The research results clearly show that interdisciplinary and intercultural education programmes will benefit the future of design, engineering and management in the textile industry: Innovations in the advanced textile digital world will be driven by the identification of textile talents (and trailblazers). As previously mentioned, these talents are necessary for a transformation and an abstract understanding of the key factors driving the Fourth Industrial Revolution. Also, they can facilitate the development of satisfactory working conditions – based on positive interdisciplinary and cross-cultural teamwork, as well as inter- and transcultural education programmes. Please compare this to and take into account the need for developed 'parameters' as outlined above, in more detail within the following chapters 4, 6 and 7.

Looking back to the 1960s, when the Braunschweig University of Art and other universities of arts began teaching design education, Roland Barthes' study of the 'Fashion System' was also attempting to find and generate a special field of 'design theory'. He, as a philosopher, said:

'My main intention has been to reconstitute step by step a system of meaning, [...] to reconstitute the semantics of actual fashion. [...] The object of analysis [...] is a true code, even though it is always only 'spoken'. Hence, this study actually addresses neither clothing nor language but the 'translation', so to speak, of one into the other, insofar as [...] a system of signs.'[39] (Barthes, R. 1967)

Along those same lines, design theory represents systems of signs and behaviour. This concept was outlined in 2013, and has been proven again several times since then – in cross-cultural contexts at lectures and project-based workshops, and as described in the ongoing chapters (particularly chapter 4).

In this context, it is important to understand the difference between the power of product design within a three-dimensional object, compared to fashion or textiles representatively used as a medium for creating a design system. This means a 'design system' with the typical characteristics of an object, in the same way that philosophers in ancient times had described the Roman style of clothing. Or, like in 1955 when the 'New Look' conveyed a new symbolic meaning.

The field study 'Textile Traces in Germany' revealed to us a new way of thinking about design systems: if we compare the 'Little Black Dress' – a classic design – to the 'Black Box' in product design, it becomes clear that you have to understand how to analyse, interpret and reflect on the semantic meaning of classical objects. There are different levels and methods for looking at art and design systems. This can be done from a historical perspective or an anthropological perspective, as well as by looking at the sociological agenda (see Fletcher, K). The semantics can be taken into consideration in order to analyse the meaning of artificial codes, and the ethnological perspective of human behaviour in 'using' things, sculptures, services and concepts.

[39] Barthes, Roland, 1983 (1967), The fashion system, translated from the French by Ward, Metthew and Howard, Richard, London: Vintage Books, p. 8, foreword f.; Compare: French original: La Système de la Mode, 1967, p. 8.

As described in chapter 2 by Wachs, a design theorist with an education background in crafts and industrial design, she has applied a new system in interpreting design codes in 'fashion theory' (Wachs, M.-E., since 2010): 'from cut to context – a six-level analysis for interpreting and creating objects and concepts'.

In the research study 'Textile Traces in Germany', she translated this model into 'textile theory' and harvested the benefits from the relationship between textiles and design engineering as applied in industrial design, for on-going university courses (see Wachs, M.-E., in bachelor and master courses in Germany since 2017 until today, and an international manuscript used at RCA, London – guest lecture 2019). This six-level analysis, shows how we can profit from looking at design history to find a sustainable model for designing the future – in a semantic way – and, at the same time, benefit from thinking about designing in production environments and the *design* of the working conditions. The semantic meaning of product languages can be adapted to state-of-the-art design engineering conditions, which relate to the textile design engineering education – today and tomorrow!

'is it time for new >forms of living<?..., based on thinking ahead with new systems of theories – Textile Theory and Theory of Creation, latters reflects the currend need for designing new habits for a sustainable future TOGETHER'

(Wachs, 2022)

3.1.6 Sustainable thinking and acting in textile design engineering & design education: a question of cultural values and a passion for regional (working) conditions

'At this school the work, created by hand, shall be ennobled to the rank of True Culture of Form, which must firstly lay its foundations in the Braunschweig and Lower Saxony regions, so that it may form values that are created in cooperation with other art colleges, and shall later evolve into economic values'. [40]

These words spoken in 1953 by Karl Wollermann, who was the first director of the first school for applied arts (Werkkunstschule) in Braunschweig, are a culmination of those spoken by several design educators in Germany since World War II. Rebuilding with one's own hands: this generated a new spirit and passion for innovative forms and creations. They incorporated values of the past, yet it was also **time for new forms, and time to live with 'objects' – forms, materials and product languages –** that influenced the everyday behaviours and life paths. This rebuilding represents not only the great narrative value of an individual's textile heritage within the cultural memory (see Assmann, J.), that could be represented by an object that belonged to their grandmother or was made by their grandfather. It is also a collective heritage, which revives our social values and shows us the state-of-the-art technological developments of former times, from which we draw additional benefit.

At the same time, this memorable textile story inspires us to rethink the conditions of former times, including rituals (see Assmann, A., 2018 (2009), Auf dem Weg zu einer europäischen Gedächtniskultur), as well as representative objects, products and processes – and places.

It also gives us the opportunity to project the needs of tomorrow and formulate the framework in which the next valuable heritage is to be created. Treasured textile artefacts are proof of the existence of corporate identities as manifested by people, things and regions. Therefore, the study 'Textile Traces in Germany' focusses on textile identities and future 'sustainable textile traces'.

In the future, it is likely that everybody's education will be expressed through an attitude – a question of cultural value and a passion for implementing creative design processes, with a 'special habitus' in sustainability and the incorporation of sustainable thinking and acting:

'Design is connecting European textile places and generations for the digital textile world for tomorrow – sustainable textile (cross-)cultural value is generating profit for people and our planet, nothing else.' (Wachs, M.-E., 2018, at conference 'Textile and Place', Manchester School of Art – Manchester Metropolitan University)

[40] Wollermann, K., 1953, Werkkunstschule Braunschweig im Jahre 1953, see chapter 2.

3.2 Driver of sustainable (industrial) design culture – the 'design shift'

3.2.1 What does high-quality product design mean?

Keywords
industrial design as a driver of sustainability, cultural behaviour – cultural changes and changes in design methods – design turnaround, the need for cross-scenario thinking in sustainable design education, sustainable thinking for complex and advanced industrial solutions; correlation between how people identify with objects, regions, changes in working conditions – reverse design: design shift.

The high-quality design of products is a consequence of sustainable design, and the investigation of origins, demands and additive design qualities – while simultaneously maintaining a focus on responsible cultural behaviour and material conduct. It is necessary to think, manufacture and invest in a sustainable manner, in order to produce responsibly and develop circular concept models for economic cycles. In addition, cultural education is the greatest investment in sustainable economies. All these factors are once again founded on our industrial heritage, and, of course, on the broad individuals produce over the course of their lifetime, as well as people showing great passion for design, it is expressing a commitment to saving the environment (see Wachs, M.-E., 2008)[41]. In terms of the digital revolution of the 21st century, we have to consider a 'design shift' (Wachs, M.-E., 2018, Conference Textile and Place, Manchester School of Art), comparable to the cultural 'turns' (see Bachmann-Medick, D., 2007)[42] of the 20th century, on the one hand. While, on the other hand, we have to take a look at the changing working conditions in industrial design engineering cycles – and the resulting **consequences for design and production processes, as well as education, of course. The following chapter discusses the impact of inter-scenario design**, related to the creative process and the impact of individual design power.

The term 'industrial design' – as described by the German industrial designer Dieter Rams –, first designates the functional needs from the consumer's perspective; human beings and how they deal WITH products. This is viewed in relation to product design, as identified by the anthropologist Michael B. Schiffer: *'The concept of life history is known in a variety of fields including engineering […] in which product design models are broken down into the major steps, such as procurement, manufacture, and use, to identify performance requirements for a technology's various activities.'* (Skibo, J.M. and Schiffer, M.B., 2009)[43]

Secondly, the industrial design process is compared to the conventional design methods of sketching, and the phenomena of so-called 'design driven by technology' or 'design driven by material'. Form and material in relation to product usage, are in line with production possibilities.

[41] Wachs, Marina-Elena, 2008, Material Mind, Dr. Kovač.

[42] Bachmann-Medick, Doris, 2007, Cultural Turns, Rowohlt.

[43] Skibo, James M., Schiffer, Michael Brain, 2009, People and Things – A Behavioral Approach to Material Culture, Springer, p. 9.

For example, the Braun razor, designed by Dieter Rams in the 1950s, is the result of a linear economy: Sketching a razor created the parameters for production, while the marketing strategy and retail possibilities were not considered until after having the industrial tools and the moulding dies made of metal.[44] Industrialisation created the basis for design management and a linear, sequential planning process. After the sustainability revolution at the end of the 1990s, when awareness grew regarding the lack of materials and that natural resources, such as mineral oil, were finite, the pressure to develop the circular economy – that is described below – became evident. Within all three pillars of the defined subject of 'sustainability', reaching beyond just materials, is not an invention of the 21st century, rather than the Brundtland report (1987) and the triple bottom line by John Elkington came to mind since the 1970s movements with the help of environmental engagement by the Club of Rome (found in 1968) for example. However, with regard to the economy and the interest of brand managers – as well as stakeholders, of course – it is proving hard to integrate into all business levels.

We have to regard the early design trigger points of the industrial design culture – rooted in the so-called linear economy – as a long-lasting standard, which has changed as a result of 'design drivers' for sustainable industrial design that are currently promoting the more favourable circular economy. As the digital revolution is now pushing us to new paradigms – forcing new process parameters and interlinked production teams – we have to think about redefining the field of 'industry', and respectively a new field of industrial design engineering and modified design methods, of course. As a result, the following question arises: How can we create a new term or expression for the word 'industry' that represents current phenomena and changes – indicating sustainability and a 'decentralised economy' at the same time (– as sign of the paradigm shift)?

We have to ask ourselves what is driving design for the next era of digital design generations, particularly when changing production conditions will be accompanied by new design drivers that do not follow any hierarchical structure. Ultimately, the potential impact will be discussed within the framework of a 'design shift' with regard to design and 'production' processes – and of course design engineering education. The following parameters are important considerations in this design process, for the industry and universities.

— changes in technological processes in relation to life patterns and their effects on identification
— media and cultural behaviour concerning objects as the driving factors for developments in design and societal innovations
— comparable design studies within the scope of historical, sociological and anthropological evaluation
— the correlation between design and engineering in the future
— sustainable education and working conditions as factors that influence changes in (and loss of) manual skills, and the need for analogue AND digitally-based fields of educational study.

[44] See exhibition 'BRAUN 100' at Bröhan Museum Berlin, 04/2021 to 08/2021 https://www.broehan-museum.de/ausstellung/braun-100/)

'changes in production processes are the result of ongoing industrialisation and change material and cultural behaviour'

(Wachs, 2020)

With the help of our cultural design heritage, as discovered in comparable European studies, the significance of three-dimensional collections can be discussed in relation to industrial production processes. Objects not only represent the high narrative value of our design heritage within the cultural mindset of the individual (see Assmann, J., 1992, Das Kulturelle Gedächtnis; ibid: see Müller-Fink, W., 2010, Kulturtheorie). It is also their collective design heritage that restores our social values and shows us the state-of-the-art technical developments of the past, from which we can benefit. This heritage inspires us to think about the conditions of the past, its rituals and the pioneers of the time. In addition, it gives us the chance to envision the needs of tomorrow's design processes.

Valuable artefacts provide evidence of the identities of people, things and regions. Such evidence reveals the typical lives as they relate to the industrial working conditions and individual skills. They are the testimony of our social and economic history, showing the cultural and material behaviour that impact cultural studies and, naturally, design studies.

However, when thinking about our textile heritage, it is not limited to woven fabrics made at home in rural areas, as was the case until the establishment of industrialised manufacturing cities. This heritage evolved with the invention of innovative machines, like the Spinning Jenny (James Hargreaves), in 1764, and the first steam engine by James Watt. Such inventions are what drove the next phase of civilisation's evolution. More than ever before, the new influential cultural aspects had an immense impact on so many individual lives – particularly those of women – which was a result of rapidly expanding cities. The driving force for the industry in countries such as the United Kingdom not only came from the inventions and innovators, but also the workforce: for example, a million workers in and around the city of Manchester, in 1800 (see Bremm, K.-J., 2014, p. 12 et al). Other industrial textile cities and centres with high technological expertise subsequently evolved in Europe, such as in Lyon in France, St. Gallen in Switzerland and the Lower Rhine region of Germany. This was a time when high-tech objects were being produced on an industrial level, and the effect of innovation gave rise to a new value of objects. Older generations placed a much higher value on the handcrafted things, viewing them as durable, long-lasting goods that were evidence of a common family heritage. Other people who were living in the cities – particularly younger generations – embraced the new richness of possibilities and 'industrial = cultural' goods. They considered the new industrial goods to be an improvement on the quality of the products – and of course of life. Material-based products were driven by the innovators' passion for industrialisation, as well as new technology. As such, the first acclaimed designers were held in high esteem in European markets.

[45] See Funken, Christiane, 2016, Sheconomy, Bertelsmann, p. 18 et al.

History's path (not only regarding industrially designed and produced products, in a linear project management) continued in this direction through to the late 1980s. At this time new designs began to be developed that served more as service solutions, evolving as a consequence of experiences drawn from artificial intelligence rather than product-centred solutions. These design concepts derived from automated production processes and the emerging information technology (IT). Last but not least, design concepts were also developed in relation to the internet as a working tool with consequences to the process.

The sociologist Christiane Funken described how there is a link between a product-based economy and an economy of knowledge (see 'Wissensökonomie' in: Funken, C., 2016) [45]. This is illustrated by the fact that most first-world companies, such as Henkel, Vorwerk and others, are no longer producing products based solely on materials, such as cleaners, electrical goods and carpets. Instead, they are producing 'concepts' with idea-based solution strategies. Nowadays, we have all switched to talking more about production services and design concepts – rather than designing objects. The change in cultural behaviours with regard to objects, expresses cultural transformation at the end of industrialisation.

'changing working conditions result in changes in the paths people take in life and their careers, as well as how people identify with regions, companies, their work and society – thus culture'

(Wachs, 2018)

Changes in technological processes have consequences and result in changes in the way people live and societal patterns, thus bringing into question people's self-identity and the way in which they identify with brands, disciplines or regions. Identification has previously been assimilated to local production sites and headquarters, which served as memorable places that also shaped morality. Identification with a product and a brand was frequently cultivated by the family's working conditions (case study of silk production in Krefeld, the Lower Rhine region, Germany, e.g. Verseidag GmbH; ibid: Mönchengladbach, Germany, Tuchfabrik Willy Schmitz GmbH & Co. Kg.). The success of a brand's product is linked to the technological know-how and the experts – the impassioned talents who are the pioneers in their discipline – as well as the involvement and loyalty of the workers, designers and inventors. (This fact is proven by the talent Franziska Poddig in the years 2020/21 again. (see IMAGES 02, p. 118 – 121)

It follows, as is evident in the past, that corporate espionage and sabotage play a significant role. We can see how corporate espionage has been occurring since the beginning of the 14th century, when skilled silk workers left Lucca, Italy (see Epstein, St. A., 2009, An Economic and Social History of Later Medieval Europe, 1000-1500, p. 212 ff.), and were regarded as veritable goldmines by their European competitors. Textile espionage was widespread between Great Britain and Germany after WW II, during the 1950s and 1960s (see Bolenz, E. et al. 2010, Die andere Schönheit – Industriekultur in Nordrhein-Westfalen). To combat this phenomena in eralier times, the British Government invented the seal 'Made in Germany' at the end of the 19th century, which was designed to be a mark of poor quality and shame. However, this attempt at sabotage actually backfired and became a mark of high quality.

To conclude the aspect of brand identification, a location is not only a place of memories but of morality. Thus, looking ahead to a post-digital time in the future, in which the 'working place' becomes physically intangible, **how can one create a memorable place to work and simultaneously a place that reflects an ethical corporate culture**? This question arises when attempting to conceive a 'new place for thinkers, workers, and designers', as is currently being developed through 'interlinked learning landscapes' within universities and companies.

After all, this stands in contrast to the conventional approach in which technology and the demand for products have mainly depended on the principle of progress, and secondly on the principle of the unity of knowledge. These factors became the focus as early as the scientific revolution in the 16th and 17th centuries, continuing during the industrial revolution in the 18th and 19th centuries (see Channell, D.F., 2017)[46]. The belief and trust in technology allowed it to emerge as the forerunner over the course of the 20th century and into the 21st century, until recently. Now, the craftsman's work, as human capital, has begun to compete with the power of marketing tools as a commercial utility (see Epstein, St. A., 2009). The epochal break with the postmodern perspective, to argue a new technoscience (Channell, D.F., et al., 2017) with an equally new post-academic scientific mindset, is linked to markets and human capital tools. The commercial demand for new methods of producing and retailing in a digitally dominated world correlates with the demands of humans who are fascinated by the new digital possibilities.

Does that mean that the belief in progress and technology is being substituted by these new possibilities and scenarios? Today, working spaces are 'homeless' – dematerialised. We can work at home for global enterprises, which are producing goods in locations to which we have no attachment or awareness, and have no localised identity that is connected to a brand's flagship or a headquarters. Simultaneously and in contrast to this, the fashion and textile companies are proclaiming great product value by providing high-quality products that are crafted based on long-term experience and tradition-based competencies. Hereby, the incorporation of heritage is the most fundamental factor (see Bühler, H., et all, 2009, Tradition kommunizieren – Das Handbuch der Communication Heritage).

What will the 'non-material textile heritage' of the future be and what is the purpose of this new world of technosciences? The parameters of the economy are time and space, as evidenced by another retrospective look at innovation in the textile industry in the year 1733. At this time, John Kay from Bury in the north of England crafted the flying shuttle which was regarded as THE key innovation of the industrial revolution (see Liedtke,

---

46
See Channell, D. F., 2017, A History of Technoscience – Erasing the Boundaries between Science and Technology, p. 256 ff.

R., 2010, Die Industrielle Revolution). The fact is that this machine – this innovation – sped up the output of the mechanical weaving looms, questioning the parameters of society's values in manufacturing and wealth accumulation. It gave expression to the enduring and pervasive high-tech culture, with which human beings serving as human capital began to identify – 'identification' as important element for motivation to act.

Even today, **the identification of employees continues to depend on the social and environmental responsibility of enterprises and their entrepreneurs** – and likely in the future will.

Yet, corporations and entrepreneurs are not the only influential bodies. At the beginning of the **21st century, the 'entrepreneurial universities'** (Channell, D.F., 2017) also began to play a role, with regard to post academic science. Science is no longer a defining factor of the innerworkings of the university, as the university context has changed and is more fluid today: the way science is organised, managed and performed (see Hinterhuber, H., 2011, ibid: Channell, D. F., 2017, ibid: Meffert, H., 2015) and how expertise is obtained, managed and organised, is different today than it used to be, due to global transformations. The changes in the workplace and societal identification (see Funken, C., 2016; ibid: Liedtke, 2012, ibid: Sandberg, p., 2013), as well as in design and engineering education, are accompanied by a new science management system (see, Fried, J., 2001, ibid: in special Lloyd, G., 2001), and opportunities found in best-practice models. These changes stand in relationship to, and are influenced by different stakeholders interests. (see here graphic 06, p. 169: 'Interplay of different stakeholders (interests)', M.-E. Wachs, 2021)

3.2.2 The correlation between design and engineering

The correlation between design and engineering is traditionally (and ritually) based on manual sketching with reference firstly to the term 'design' and the analogue activity. This has been the case since Leonardo Da Vinci sketched his flying machines during the Renaissance period, for example. This was a method of communicating one's ideas about machines and the production conditions of the future. Secondly, design and engineering both developed with the aid of technical sketches, by using digital tools as well as new production possibilities, to depict design as an object: an item of furniture, a lamp, a cooking appliance, a textile cover. Beside this, German engineers have had a different understanding of the terms 'design' and 'designer' up until today. For designers, these terms mean using the sketch to create the tool and the machine to produce the design object. Yet, in Germany, a hierarchical distinction is still being made between engineers and designers – thereby especially elevating engineers, who also promote this notion.

Nowadays, when creating in the automotive, textile, or fashion industry, as well as other disciplines around the world, we utilise different analogue and digital methods, as well as combinations like analogous digitalised methods for computer aided design (CAD) processes. In contrast to robots, we, as human beings, need and profit from the **analogous experience of a 1:1** model, or smaller scale models and prototypes. These help us optimise and perfect products. The combination of 3-D printers and other small-scale machines, **once again highlights the prevailing economic factors: time and space**. In the 21st century, the 'design engineer **driver**' and the reality of the world's limited **resources** are now also key factors, as we work towards sustainability and circular economies. However, creating circular economies and the commitment to profit are not the only underlying trigger points for new designs, products and services. Since John Elkington's Triple Bottom Line, focusing on 'Profit, People, Planet', a new perspective has come into focus, which takes on far-reaching responsibility by means of a holistic approach towards sustainability (see Meffert, H., et al., 2015). This means implementing non-hierarchical structures for developing, producing and retailing objects and services in the future, for industrial production and management processes.

It also implies that we are not looking at a future with a linear or circular economy, but rather a complex system with non-hierarchical economies, in which all participants can influence the industrial process: the entrepreneur, the worker, the designer, the general public, the expert, the scientist and even the child. This gives us an advantage when creating the future. Although there are some disadvantages, the benefits of implementing such structures far outweigh the negatives when creating products for the future ultimately, this approach will create new frameworks for design and engineering education and – **after the linear and circular economic models – it will provoke a decentralized, 'star' based economy**, where for example children and entrepreneurs are 'sketching the future together'.

3.2.3 The design turnaround, the digital revolution and de-industrialisation: How will the design shift impact design education?

'De-industrialisation' (Bärtschi, H.-P., 1998) and digitisation in Industry 4.0 are the dominant parameters for the driving forces in industry today. This applies to both brands belonging to textile-production companies, as well as those in the metal processing and production industry. We have to focus on **changes in the significance** of construction parameters and design methods. Design thinking is a matter of subsequent dematerialisation on both sides. When creating and producing in design engineering, it is important to remember that interdisciplinary, cross-cultural, non-hierarchical working teams are the future. The transformation of production processes into the post-industrial era a digitalised world will be directly connected to cultural behaviour. Ergo, it is linked to the behavioural framework of human beings.

Technology transfer is usually interpreted within the following context: the transfer of a technology from one societal or cultural context to another, and from a higher developed to a less-developed space (see Skibo, J.M. and Schiffer, M.B., 2009). The cultural context we create begins within the university itself, putting sciences into a theoretical context. The remaining framework is created externally, by establishing cooperative projects with companies, experts and talents in European countries. With regard to our textile design heritage, we have so-called 'lieux de mémoires' (Nora, P., 1980), places of memory, which can serve as resources for transference. Here, experts, teachers, and talents are in the process of collecting,

QR Code 01

QR Code 02

and creating 'knowledge banks and archives of knowledge' (Wachs, M.-E. 2019, DHS Congress Northumbria University, Newcastle), which are also attracting other impassioned people. **Thus, the advanced (scientific) research commitment continues – as in the past – in** Mönchengladbach, Germany; Borås, Sweden; Manchester, Great Britain; and Herning, Denmark. All these places are not only collecting objects, but also testimonials of good form and crafted design objects – and knowledge! To protect this immaterial legacy (in the form of knowledge), we must manage it responsibly because it is the resource from which new inventions will evolve.

Following this supposition, the interdisciplinary and intercultural design projects 'Textile Archisculpture' (see QR Code 01 and 02) and 'Textile Traces in Germany', by Wachs, promote a circle of mixed teams bringing together European talents and experts from textile centres that are home to great textile heritages in each region of Great Britain, Sweden and Germany – recently of Denmark and Italy within new projects in 2021/22.

Young European students and experts have taken the opportunity to use knowledge transference to create textile visions in hybrid spaces with dynamic design studies – in both analogue and digital formats. They have used conventional design methods, e.g. field studies, but also implemented modern design methods: design thinking, transformation design and design scenario stretching. Design studies included 'KURT – simple, solid, smart: the textile heritage of Krefeld inspiring sports couture and textile architecture', by K. Amprazi; 'Dampfenergy – relax and stretch: a fashionable yoga machine', by K. Grobheiser, fashion and textile designs like 'Krefeld reloaded', by F. Poddig and 'BRICKED' by E. Esser and 'Illuminated Brick' by Lena Eiche (see in digital catalogue to the project 'ArchiSculpture', QR Code 02), to mention a few. These design studies focused on advanced sustainable textile design solutions for fashion and archisculptural design, by analysing the textile heritage at places of memory and implementing high textile engineering standards. The research that the talents did on textile's heritage gave them a new consciousness of values for the (design) future.

'sustainable education environment means human centered working conditions – including: coaching, mentoring and initial teacher education tools –, these factors are influencing the needs for analogue AND digital education-based programmes'

(Wachs, 2021)

The digital revolution at the beginning of the 21st century is only one of the causes of a 'design shift' (Wachs, M.-E., 2018, London, EPDE Conference) and the reason we need more sustainable thinking in design and in design education. The change of industrial production processes correlates with a change in design methods. The design shift also questions the evaluation of design objects and the parameters of the circular economy, as these are proof of the change in correlations between different corporate identities – of people, things and regions. This development generates potential for new drivers of sustainable industrial design systems in a non-hierarchical structure. At least for the next decade, discussions will need to have multifaceted approach, by taking members of the industry, universities, and the younger generations – even those in pre-school – into consideration. Also, new collections and archives, in which heritage-based knowledge is collected, will come into play as outlined in chapter 7 more in detail.

The consequences for production sites and industrial conditions in the future are going to be influenced by the way in which people identify with values, such as sustainability, product longevity, our planet's resources, and ethical values which will impact our lives. Communicating heritage as a management tool goes hand in hand with communicating tradition, both of which are linked to life experiences, places to remember and the paths individuals take in life. These will remain **socially relevant within brand management and indeed entrepreneurial universities as we move forward**. In the future, after de-industrialisation, which will likely attract people to certain locations – not only for work – will mean **continuing to keep human skills alive and in use, as well as creating new 'rituals' for tomorrow – thus, mediating visionary learning landscapes**.

Design studies are currently focussing on textile and engineering design identities in relation to education and workplaces with regard to history and the future. When it comes specifically to 'sustainable design studies' for the future, the thesis puts forward the case that the following is necessary: **a different design education at (pre-)school, more effective intercultural connections between universities and teams on the European continent, and the creation of a mobile task force of talented design engineers**. This is what is needed to transition to a sustainable way of thinking and acting in design, enabling the understanding that sustainable design creates sustainable long-lasting living conditions.

Within the framework of resilient living, the technoscience perspective is only one possible way of looking at the next drivers in design and engineering. The new design generation is sketching a sustainable design culture for a digitised and cross-culturally connected society – yet, one which will still be localised in terms of the people, things and stories of our future cultural heritage. They are the drivers for sustainable design culture, resulting from the design shift, who will profit from a very well-managed 'Knowledge Bank' (Wachs, M.-E., 2019). Ultimately, this will result in a new form of 'management of science's.

PRELIMINARY APPRAISAL 01

(by Wachs, M.-E.)

Common and differing characteristics in **product** and **textile** design engineering – the business of 'industrial design' in the past and the *'design shift'*

As we have seen in chapter 3, textile design engineering is not only about a unique 'material' story, showing that textile design history 'was' related to places, textile products and textile heritage – but also to textile 'nerds' and people with a passion for the business of textiles.

A focus on the design education heritage in Germany, not only the Bauhaus imprint, but also the characteristics of the hands-on design of Germany's Werkkunstschulen and the legacy of 'industrial design', analysed the current day relationship between the industry, and designed products and concepts: As we have seen in chapter 2 and 3.1 (in particular in 3.1.6), form and function will have to focus on sustainably designed solutions, to create the engineering heritage of tomorrow. This will be dependent on holistic design engineering understanding and education programmes, as well as a thinking process that revolves around sustainability.

Compared to Germany, Europe demonstrates similar design processes and a similar emphasis on product language – which is rooted in a comparable, recognisable background in design theory and an understanding for the other partners – it was and is the foundation of tomorrow's future in the post digital era, in keeping with the design shift.

Thinking and acting together in a sustainable design engineering landscape requires empathy and the willingness to integrate 'otherness' and the motivation to act. This scenario formulates the European design engineering perspectives.

For Germany, the wish can be expressed, to return to placing a higher value on art and artistic skill mediation at school, which will lead to the long-awaited and true acceptance of 'design' – on the one hand as a 'course' and shift in attitude at school, and on the other hand an attitude shift within the society.

'Design' means, 'sketching the future' in a mindful but corporeal connected, and multi-perspective learning landscape. 'Design' means, conceptualising the future through interactive and integrative hands-on and digital spaces. 'Design' means, connecting people and helping people understand objects and living concepts through a holistic view: This begins with ergonomically designed solutions to everyday problems, solutions that fulfil both the aesthetic and ethical demands on products, or the complex design system(s), as well as the highest demands on all three pillars of sustainability.

All these statements about what 'design' means hold true for all design engineering methods, regardless of whether these relate to designing with textiles, woods, CAD-systems, AI (Artificial Intelligence), or humans building artifacts with their hands. They even apply to the way things are built in nature, and the method of 'thinking through drawing', which will be looked to more closely in the chapters that follow.

Yet, in order to understand design – and the business of design – you have to create a 'manual' for reading and developing an understanding of design, as well as how to handle the various media codes of design. Manual skills, like building with sand or sketching, generates self-expression and serves as an experiential tool (e.g. ductus) and an opportunity to create 'codes' – languages that help us understand, read and interpret abstract phenomena in complex design engineering processes. This will be the most challenging demand on human beings in the future – and not only in design engineering or textile engineering!

In the past, German design meant, 'industry' and design engineering education were related to 'product language'. Yet, in the future, German design will be related to a European understanding of design systems and Western design models that promote a more interlinked complex act of designing – as best-practice models will exemplary show you in the following chapter IMAGES 02, new talents by textile design engineers.

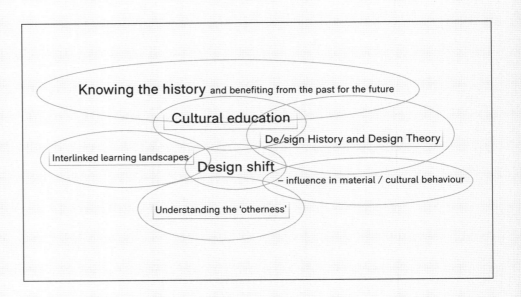

Graphic 03
Knowing / understanding the past for designing future behaviour – the design shift, with these core elements, M.-E. Wachs, 2021

IMAGES 02

Textile design drivers' Textile Engineers' success today

In the beginning of the third decade of the 21st century, the state of art of the developed discipline of 'Textile Engineering' is an important creative business field as part of the design domain. The positions and 'IMAGES', of the textile engineers new talents in the following, gives an insight in the analogue and digital interlinked design spaces for the future – for the complex problem solver our industries and societies need, and which is beneficial for our planet to build sustainably and with holistically educated mind set.

Thanks to the new talents for the precious time we created together, and let us mentoring the next experts in this way, that will be proved to be a worthwhile investment in the next generations in designing good. Please enjoy the following design engineering projects focusing on textile sustainable solutions in different applied design fields: product, interior, exterior, and fashion design and apparel, architecture, lighting design, communication design, material design. We would like to invite you, please to groove into the design / process via the video via QR codes… a wonderful addition!

Theresa Scholl
Elise Esser
Gesa Balbig
Franziska Poddig
Lydia Petersen
Katharina Grobheiser
Melissa Grustat

Theresa Scholl

Color & Trim Designer
Volkswagen Nutzfahrzeuge

Theresa Scholl is a bespoken tailor and textile designer with a Masters degree from Hochschule Niederrhein. During her studies she created new textile materials such as smart textiles, textiles for lighting design and paper textiles. Additionally, she focused on the design theoretical investigation of textiles in architecture. After having had some touch points with the fashion industry, interior design and lighting design, she is currently working as a Colour & Trim designer in the automotive sector.

Contact
scholl.theresa@gmail.com
www.theresascholl.de

A
Screenshot program Ecotect: Calculation and visualisation of the daylight impact on 'Festland' with its surrounding building; Image by Theresa Scholl for Ulrike Brandi GmbH (2017).

B
Architectural sketch by Theresa Scholl; Image by Theresa Scholl (2018).

C
Detail: Knitted curtain catching the sunlight, Design Theresa Scholl; Image by Theresa Scholl (2018).

Diploma 2018
'a textile sense of light'

'a textile sense of light' – this master thesis investigates the two disciplines, textile design and lighting design in architecture.

Technical planners often limit textiles on their decorative aspects and therefore, fail to consider the high potential of textiles in their construction projects. However, textiles can do much more than being decorative. Together with the Ulrike Brandi Licht GmbH, Theresa Scholl proved this by developing a modular concept for textile design solutions for architectural lighting design: Textile daylight systems are able to manage impact of daylight in a building through regulation of sunrays and shading. Based on an analysis of the building's daylight situation and a matrix for textile design engineering, unique textile systems can be created customised to the building and its intended use. This concept was applied to the building project 'Festland' of Hamburg Leuchtfeuer FESTLAND GmbH. Now, colourful textiles with different transparency do not only help to control the daylight impact in the building, they appear playful and underline the inviting character of this assisted living project.

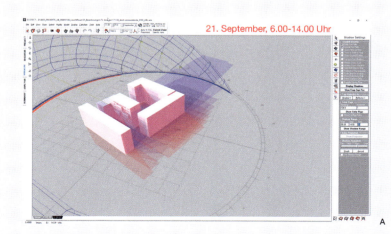

A

B

C D

Theresa Scholl

E

D — G
Project Festland: Colourful, semi-transparent curtains control the daylight impact and create a playful and inviting meeting zone in the corridor, shape a warm atmosphere in the evening inside and create a poetically light language by the view onto the fassade by night. Design Theresa Scholl, pictures by Ulrike Brandi GmbH, 2021, with friendly authorization by Hamburg Leuchtfeuer Festland gGmbH.

F

G

Elise Esser

M.Sc. Textile Engineering

Contact
kontakt@elise-esser.de
www.elise-esser.de

A
Lighted biomaterial with fine slats and detailed leaf structures.
Elise Esser, 2020

B
Raw ginkgo leaves in various structures as main component for the development and production of the biomaterial.
Elise Esser, 2020

**Diploma 2020
'Organic Shades – The natural lighting concept' and designed bio synthetic material 'Ginoja'**

The earth's resources are getting rarer and our consumption is getting higher. The way we deal with nature and consumer goods needs a revolution.

The master thesis 'Organic Shades – the natural lighting concept' reflects the potential of previously unusable natural goods to create new and high-quality materials and objects. The aim is to use the human resources of our intelligence and creativity to create something new and shape the future. Leaves from Ginkgo tree build the base of the material research. The collected leaves were transformed into a flexible and translucent leather like material. The newly developed biomaterial 'Ginoja' establishes sustainability in a luxury way to promote a more conscious approach to the environment. The lamp 'Organic Shades' made from Ginoja offers a possibility to use the material and light up its naturalness to draw attention to the special characteristics and diversity of nature.

Use your creativity – see the nature's potential – act sustainable and wisely. The only natural resources we should draw from are our inventiveness and brainpower.

A

B

Elise Esser

C
Organic Shades light
wide format in wooden base.
Elise Esser, 2020

D
Biomaterial experiments with
laser cutted shapes. Rolled
piece with geometrical pattern.
Elise Esser, 2020

E
Ginoja translucent biomaterial
made from ginkgo leaves with
fine structures and details.
Elise Esser, 2020

D

E

Gesa Balbig

B.Sc. Design-Ingenieur Textil

Contact
gesa.balbig@googlemail.com

A
Woven Design, close-up.

B
Inspiration: Silk Wall from Archi-Union Architects, Shanghai China.
Photographer: Marc Goodwin

Design, pictures, production of the design and management of shooting and communication tools by Gesa Balbig.

Diploma 2015
Design project 'silk wall' inspired woven jaquard design 2018

Inspired by an architecture project called 'silk wall' by Archi-Union Architects in Shanghai 2010, I created this texture on the TC2 jacquard loom from Tronrud, as part of a digital weaving workshop, organized by the Textile Society of America. Fascinated by the result and 3dimensional structures, more ideas followed in collaboration with Hannes Krieger, a visual effects artist, using the program houdini. Based on his 3D simulations of various surfaces, I developed complex 2D patterns in Adobe Photoshop.

In times of digitization, the technology allowing to 'draw' in a 3dimensional space are powerful tools, but aren't able to replace the tactile experience and visual depths of the physical sample. Together with modern machinery like the TC2, used for sampling in the textile industry, digital ideas can be combined with the manual, the traditional craft. More to be demonstrated and to be seen within the movie, follow the QR code attached.

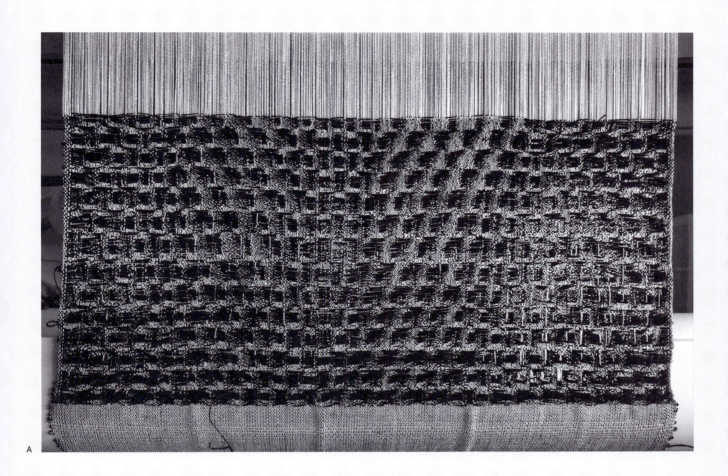

A

B

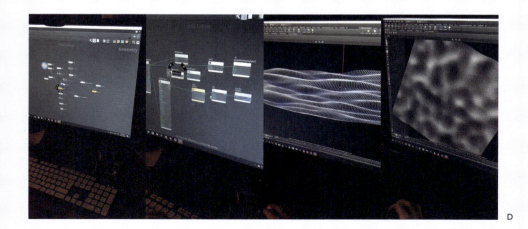

C
Woven Design on TC2.

D
Houdini 3D program: Geometry network with nodes (1 + 2) and scene view with geometry object from the side and the top (3 + 4).

E
Grid Surface – Detail.

F
TC2 loom.

Gesa Balbig

Franziska Poddig

M.Sc. Textile Products Design

Contact
franziskapoddig@gmail.com

We thank Tuchfabrik Willy Schmitz GmbH & Co. KG, Germany.

A
Collection (un)genutzt – Jacket and Sitting Ball Cover

B
(un)endlich – Fabric made from rebonded yarn residues

C
(un)geahnt Inspirationsbild

D
Collage (un)genutzt

E
Collection (un)genutzt – Dress and Belt

Design, pictures, production of the outfits and managment of shooting and communication tools by Franziska Poddig.

Diploma 2021
The (im)perfection of chance as design potential – reshaping the perception of unexploited textile production ressources for an added value of the future.

Franziska Poddig's final thesis 'The (im)perfection of chance as design potential – reshaping the perception of unexploited textile production ressources for an added value of the future.' combines the disciplines of design and engineering for the development and industrial integration of innovative and sustainable processes from textile weaving waste.

The related design concept focusses on the importance of chance not only from the perspective of sociology and science, but also its potential for the creative process. Inspired by the works of Gerhard Richter, Franziska Poddig establishes in her work moments of chance on the conceptual level but beyond that she developed the 'shape driven by chance' – a strategy to form a design language based on the idea of methodical chance.

It is not enough that a design is sustainable at any given moment, it must be thought through in all aspects, not only for now, but also in all consistency for the future.

A

Franziska Poddig

B C

D

Lydia Petersen

Fashion Design Engineer
Masterstudent of Textile
Productions – Design

During the Bachelor studies of Design Engineer in Fashion I developed a profound interest for fashion illustration. Since my ERASMUS semester abroad at the Belfast School of Art I became passionate about freehand machine embroidery. Since then, I am exploring ways to combine fashion design with fashion illustration and embroidery.

Contact
lydia-petersen@gmx.de

A
Backside of the illustration
'What if women ruled the world?'

Diploma 2021
'the new woman in 1920 and 2020 – a hand stitched and freehand machine embroidery design statement' by Lydia Petersen

This illustration series addresses the image of women in the modern feminist zeitgeist. By combining hand stitches with freehand machine embroidery as well as taking into consideration the history of embroidery, a connection between the past, the present and the future is created. During the work process, analogue and automatic / digital techniques have been alternated. In the beginning, freehand sketches were made, then edited digitally with Adobe Photoshop. Those edited sketches were then printed, cut out and placed on the fabric as pattern pieces. Those pattern pieces were then sewn to the fond by using freehand machine embroidery and afterwards covered using machine sketches.

This series represents the variety of freehand machine embroidery which can be used to sketch, to hatch or to create faces. It also represents the variety of the women's fight against patriarchy and the fight of embroidery to be acknowledged and respected as an artistic medium.

Throughout the whole of the series, the hand embroidered cross stitches are continuously depicting the rebellion against patriarchy and a break with the societal expectations of women, at the same time it is a symbol for censorship.

A

B
Development of an embroidery using freehand machine embroidery.

C
First drafts for design development, 2020.

D
Creation process of the digital illustration 'What if women ruled the world?'

Pictures by Lydia Petersen

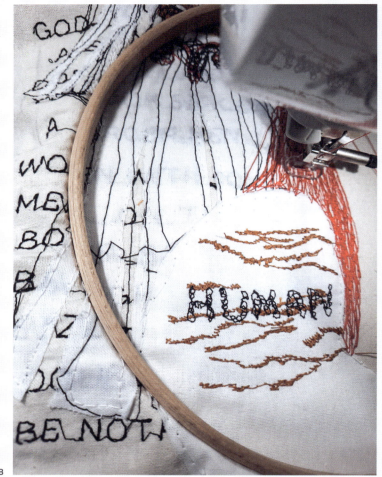

B

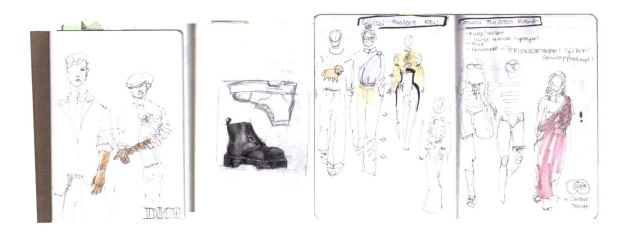

C

D

Lydia Petersen

Katharina Grobheiser

M.Sc. Textile Product Design
Textile Design Engineer

Contact
katharina.grobheiser@gmail.com
Instagram: kaykaycreative

A
FUZZY Cover with
a sewn textile surface.
Picture by Daria Buch

B
Manual sketching process
of a dryer lint. Picture by
Katharina Grobheiser, 2020

C
FUZZYS mixed with water.
Picture by Katharina Grobheiser,
2020

Design, production of the
outfit, management of shooting
and communication tools
by K. Grobheiser.

Diploma 2020
'A Fuzzy Society with Fragile Freedom create a diverse Future through Design'

Textiles are made from fibers that represent their identity. In every mechanical washing and drying process some of these fibers get lost. The Master thesis 'A Fuzzy Society with Fragile Freedom create a diverse Future through Design' reflects the personal mindfulness and the social appreciation of the free existence. This theoretical topic is united with a practical and multiple diverse design processes.
The Basis of this textile material research were fibers that have been collected by textile drying machines. An important part of this material research was a daily mindful manual sketching process. During this process Katharina Grobheiser cultivated the '8 Theses of a mindful creation' which helped her change the FUZZYS (dryer lint) identity from wasteful to a valuable textile and nontextile material and products.

A

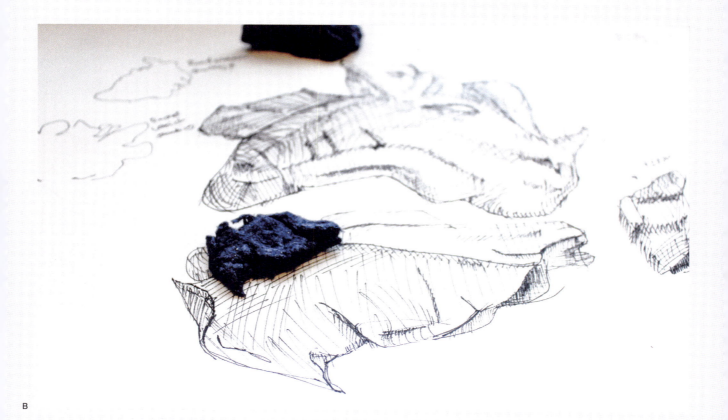

B

C

Melissa Grustat

M.Sc. Textile Products, focused on Design, B.Sc. Textile- and Clothing-technology, focused on Management, B.Sc. Design Engineer, focused on Fashion. Freelance Editor for 'AD Germany'. Five years experience in Sales, also experience in working as a designer for Istanbul based brand

Contact
melissagrustat@gmail.com
www.melissagrustat.com

**Diploma 2021
'DNair'**

'DNair' by Melissa Grustat is a thirteen-piece collection. The main component of all pieces is airbag fabric collected from various sources or sponsored by German companies. Although the origin is different, all materials would have been thrown away. The designs are meant to show what future applications are possible for this high-tech airbag fabric. All designs are linked by a pur, timeless product language, which is based on strong characteristics of 'German design', which guarantees long durability and takes care about sustainable systems – like recycling of first qualities of airbag material.

A
Seat cube made of waterproof airbag fabric that can be connected to form a large couch. Filling from car seat production waste. In addition with raincape made of waterproof airbag fabric. See picture C the raincape transformend into a transportable bag pack.

Design, pictures, production of the outfits and management of shooting and communication tools by Melissa Grustat.

C D

E

B
Cargo pants made of airbag fabric with waterproof pockets.

C
Transportable raincape made of waterproof airbag fabric.

D
Waterproof rain vests made of airbag fabric. Once in colored version.

E
Re-interpretation of 'Marcel Breuer Chair' with seat made of airbag fabric.

Melissa Grustat

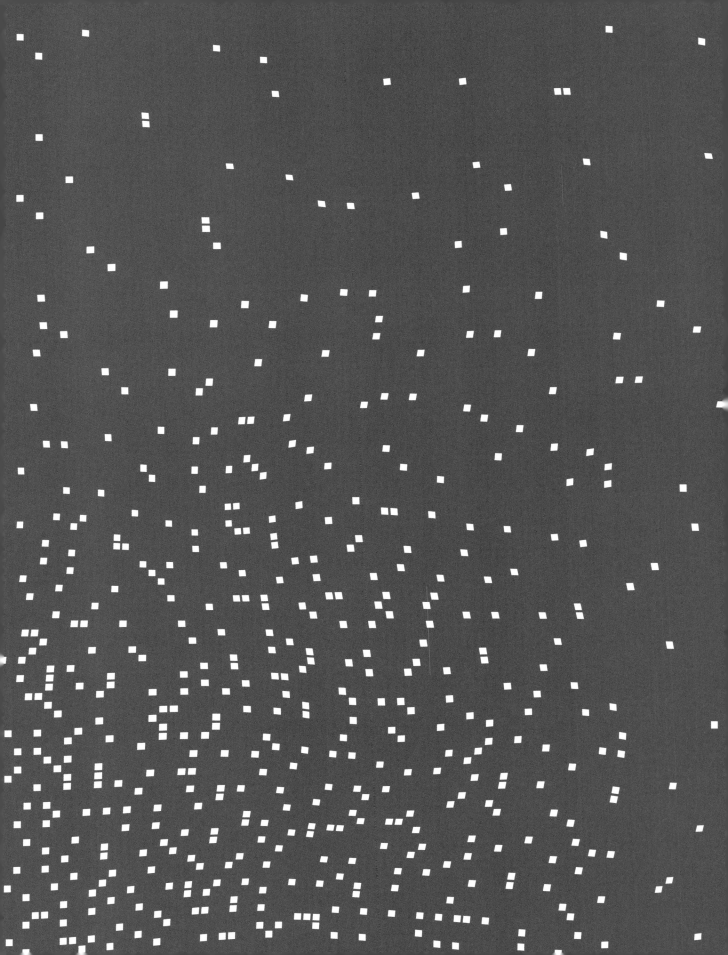

CHAPTER 04

132 — 154

Education research in practise –
comparing sustainable design
studies and 'learning landscapes
of togetherness' in Europe

4.1. Collaborative non-hierarchical design engineering and management workshops – pan European workshops in Germany

Keywords
design thinking by physically experimenting, analogue interlinked design creations, European design education; future experience education strategy: non-hierarchical cross-cultural design space, didactic development of experiments as an educational tool.

'In tomorrow's world, many new positions and professions will emerge, driven not only by the fourth industrial revolution, but also by non-technological factors such as demographic pressures, geopolitical shifts and new social and cultural norms. […] scarcity of a skilled workforce rather than the availability of capital is more likely to be the crippling limit to innovation, competitiveness and growth.'[47] According to Klaus Schwab, the development and consequences of the digital revolution's technological impact will cause us to focus on talents' capabilities and will significantly increase the importance of 'human capital'. His words are encouraging us to continue incorporating interdisciplinary projects as part of a sustainable cultural education, and to execute this in a more 'intercultural' or 'cross-cultural' manner. This development has inspired us in Germany (Dusseldorf and Mönchengladbach), as well as those in Great Britain (London) and Sweden (Lund) to organise European workshops that revolved around design creation in the year 2019.

First, the case study about European research in sustainability, a German driven European design perspective, underlines open minded creation in interlinked working spaces. A discourse and experiments focussing on Europeans in collaborations that foster a profitable 'togetherness' follow in chapter 4.2 to 4.3.

The workshop event '*TextilePop_Europeans together in SUSTAINABLE design / engineering / management*', took place in Germany, in May 2019, during the week of the European Parliament elections. It reflected Marina Wachs' design research efforts of the last ten years as a member of the Faculty of Textile and Clothing Technology at the Hochschule Niederrhein – University of Applied Sciences, in Mönchengladbach, Germany. The event was a collaboration between the NRW-Forum in Dusseldorf, the Hochschule Niederrhein – University of Applied Sciences, and European guests. Representatives came together from all possible fields of textile competencies in design, engineering and management – with a focus on SUSTAINABILITY – to collaborate in four exposition spaces. In addition, analogue and digital exhibitions of various sustainable design engineering and management solutions were presented. The non-hierarchical, interdis-

[47] Schwab, Klaus, World Economic Forum, 2016, The Fourth Industrial Revolution, Penguin, p. 44 ff.

ciplinary workshops in 2019, illustrated how the open-mindedness of the cooperation partners and alumni, who have a real interest in designing and learning together with bachelor and master students within a non-hierarchical framework, is beneficial to everyone.

The goal of design research is to communicate, connect and create together – on an interdisciplinary and cross-cultural level – to prepare for design engineering tasks and design management exercises of tomorrow's new interconnected global generation. The following topics were the focus of the research interests in the workshops:

— **sustainable** design and design management **solutions**
— **fostering a united Europe** with the help of intercultural and interdisciplinary projects in design and design management – to **prevent increasing geopolitical problems**
— to underline the significance of **cultural education** as a basis for design and design management
— to focus the challenges and **chances that the transformation of the digital** revolution provides
— the **gender equilibrium, diversity management** in design, engineering and management
— lifelong lessons taught in non-hierarchical structures – from childhood to retirement
— **'osmotic' education spaces** for individuals, industries and societies.

The results of the workshop projects revealed an extensive range of creative potential, scope of technical design, ability to comply with sustainability demands, as well as the success of collaborative engineering solutions. The workshops revealed an excellent standard of interdisciplinary networking between the experts in various scientific and design disciplines, industry partners from the Lower Rhine region, Europe and beyond, and the young talents – our human capital of tomorrow. Many solutions showed insights into the future of textiles, which may at some point be considered state-of-the-art.

Yet, TextilePop offered more that just an opportunity to foster interconnectedness. The framework of the exhibition also created 'a place to remember' and to connect in design. In addition, TextilePop tried to sketch out and research innovative tools for education programmes, as well as apply solutions for the industry in a non-hierarchical format: **from school to university, to circular industry management**.

Furthermore, the exclusive workshops, guided tours and interactive playground for children took place during the week of the European Parliament elections to underline a common sense of purpose. It highlighted the common mindset of European design heritage and the shared interest in sustainable design engineering responsibility.

A space for textile makers and design thinking workshops were an ambitious addition, serving to mediate smart sustainable solutions for the post-digital era for all generations. The exhibition was created using tangible 3-D objects in combination with films and animated pictures. In addition, background information was provided by guides during tours, hands on experiences and via digital catalogues and audio guides for each exhibition position. Working together and thinking about educational fields of study, learning about the future landscapes of European students and experts who are dedicated to SUSTAINABLE Design / Engineering / Management, or help bring to live creative landscapes and new, non-hierarchical 'European togetherness'.

It was a great honour to realise the innovative design-learning landscapes of non-hierarchical, cross-cultural workshops within the framework of the exhibition and in a collaborative space with European partners. Partnerships with cooperating enterprises and universities (Royal College of Art in London and LUND University in Sweden), have now become highly valuable. We welcomed great experts from enterprises that are the so-called 'hidden champions' of the European economy. These businesses are dedicated to developing sustainable, value-based working and living conditions, while simultaneously creating a clear perspective for the young generation and talents. The collaboration at the NRW-Forum in Dusseldorf (Germany), allowed young students and talents from the alumni pool, with backgrounds in various disciplines, to work in interactive teams with experts from various European countries and business fields, as well as teachers. They formulated a strong European statement regarding smart sustainable tasks and sketched possible advanced solutions for the post digital area, also in learning TOGETHER. In the following part, the four workshop spaces, guiding experts and characteristics in designing are described. The workshops took place in the environment of the exhibition, with the support by Hochschule Dusseldorf – exhibition design.

1. Light – Textile – Space: 'Enlightening new spaciousness and immateriality through textile', was guided by the expert in lighting planning, Dipl.-Des. Ulrike Brandi (Ulrike Brandi Licht GmbH – Hamburg, Germany), together with alumna M. Sc. Theresa Scholl (currently working as a junior designer at Volkswagen Germany)

2. Next Material: 'From Waste to Value – Material Matters' was guided by Prof. Dipl.-Des. Ellen Bendt (Hochschule Niederrhein, Germany) in cooperation with IMAT UVE Material lab (Mönchengladbach, Germany)

3. Fashion Textile / Engineering: 'Change of Role Model in DESIGNING Sustainable Warp Knitting' was guided by Dipl.-Ing. Markus Rindle (Textile Engineer and expert for sportswear and apparel in Switzerland), together with alumnus B. Sc. Leo Aha (Master student of Trade & Retail, Hochschule Niederrhein, Germany)

4. Fashion Management for the Post Digital Era: 'Fashion Management within the Scope of Sustainability in Germany' was guided by Sara Teske (Fashion Manager), supported by CEO Scott Lipinski (of Fashion Council Germany, Berlin).

QR Code 03
See abstract
by Wachs / Hall,
at EPDE conference
2019, Glasgow.

QR Code 04
See presentation as
pdf by Wachs / Hall at
EPDE 2019, Glasgow:
(more at research gate)

QR Code 05
See Proceedings as
pdf by M.-E. Wachs.

The workshop was introduced by fantastic key speakers, like Ashley Hall, from the Royal College of Art London, and his presentation about 'Tangible aesthetics in design engineering and the post digital era', as well as an inspiring interaction in 'Sketching the future ...into yours!', including a sketching tour with Charlotte Sjödell (industrial designer and teacher at Lund University, Sweden) and Siri Skillgate (industrial designer in Sweden). All three teaching experts have practical experiences as 'industrial designers', and afterwards they worked very interactively, allowing the participants to contribute to the different workshops. This can be seen in a small video of the workshop 'Textile and Light', in which a **teacher**, two master **students** and an automotive interior design **manager** with great expertise created a new solution for light in textile together – as partners on equal footing. The BA and MA students worked passionately together with design managers and leaders from the textile and fashion industry; William Franke (Head of Design, Colour and Trim at AUNDE) and Tom Wessling (Head of Design at BRAX), for example. Insights into the event can been seen in the movie and proceedings. (see QR Code 05 / QR Code 06, p. 138)

It is our goal to *'require new educational talents working seamlessly across integrated analogue and digital platforms while responding to evolving cultural needs emerging through new consumer behaviours. As geopolitical changes accelerate in Europe, new opportunities and challenges are emerging in collaborating for a profitable 'togetherness' (Sennett, R.).'* (Wachs / Hall, statement at the EPDE Conference 2019 in Glasgow). (see QR Codes 03 and 04)

Interim results of the workshops and of the innovative learning platform were qualitatively evaluated. These results have been published on various (digital) platforms, such as research gate. In addition, they were presented, discussed and published at and by the EPDE Conference – the Engineering and Product Design Education Conference – by Hall and Wachs together at Strathclyde University, Glasgow in 2019. (see QR Code 03)

Wachs's dedication has continued through workshops in Great Britain and discussions with teachers and experts in 2019 in United Kingdom and Germany and as well in July 2020 through an interactive workshop at the VRHQ lab (Virtual Reality Headquarters Hamburg), leaded by Roland Greule at Speicherstadt Hamburg, Germany. Here they discussed the interactive, immersive design process of creating and 'sketching' within virtual spaces of VR (Virtual Reality) and MR (Mixed Reality), the integration of analogous realities, of VR glasses and haptic gloves, and the vision of 'textile tabs' (Wachs), which will be developed in the future. This scenario describes the need to first work in teams characterised by different businesses and generations, and then through a combination of analogue and virtual experiences. This connected creative area, a more **haptic based sketching tool, linked** to white boards and other digital devices is to be continued – which is here part of future education strategy and to formulated in chapter 7 (see Marina-E. sketching in AR, p. 269).

So, these have been the first steps in Germany and Great Britain to greater European cooperation within **non-hierarchical learning landscapes in design engineering and management**. – We would like to encourage external partners to follow in our footsteps, by generating a

Note
Ashley Hall and Marina-E. Wachs during presentation at Glasgow: EPDE Conference, 2019, pic: M.-E. Wachs.

Picture 01
Marina-Elena Wachs and Ashley Hall (RCA) within panel talk to 'interdisciplinary interactivly designing', pic: M.-E. Wachs, 2019, Dusseldorf.

Picture 02
Impressions of the interdisciplinary workshop: 'textile and light', 2019, left: Ulrike Brandi, Charlotte Sjödell, two Master students, William Franke. pic and movie: M.-E. Wachs, 2019.

Picture 03
The 'textile buffett' – with thank to M. Zellner GmbH, https://www.zellner-textil.de – for designing with textile and light; see as well within the movie – via QR Code 05, pic: M.-E. Wachs, 2019 during the European workshop at Dusseldorf.

QR Code 06
Movie workshop 'textile and light', Dusseldorf 2019, movie rights M.-E. Wachs, with friendly autorization of the workshop participants, 2019.

more physically connected immersive digitalised process: together we must support the cultural education in Europe for the talents of the future – the experts of our future – and for the benefit of humanity and our planet around the world. In this framework, it is very important to understand the beneficial aspect **for both parties, industry and education institution, that 'design thinking' needs the process of physically experimenting**.

Very ambitious talents and experts are aiming to create a sustainable and worthwhile future, while still satisfying the needs of the Industry 4.0. But what's next? What about the enormous challenges that the equilibrium between humans and nature is being confronted with, in the face of diverse economies, geopolitical changes, social and psychological challenges, as well as the transformation of our digital world filled with digital products? Hopefully, this design research book will inspire you – consider it an invitation to accompany us through future workspaces: creative, innovative, smart, sustainable spaces in (textile) design engineering research. Here we can discuss and generate the solutions for tomorrow together. The paradigm shift that is the consequence of the 'Fourth Industrial Revolution' is taking place now, and we have the chance to create a new cultural space, and formulate a sustainable framework. We agree with Professor Klaus Schwab that **the talent, rather than capital, will represent the critical production factor. The human capital will be our most valuable asset and the driving economic factor** in the years to come. Design engineering research in Europe requires a connection on a deeper level, with new edcational models. We must discuss this issue. As such, it would be beneficial to connect German design research with yours – in Europe and beyond.

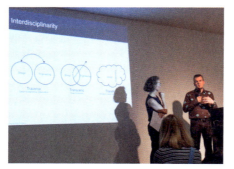

Pic. 01

Pic. 02

Pic. 03

4.2 The method of 'material-based design thinking' with the aim of 'materializing immateriality' – a pan European workshop in Great Britain

Keywords
Material based design thinking, to foster European design understanding in design education; non-hierarchical; cross-cultural design space, didactic development in education experiments, materialising immateriality, material-based thinking, analogous and digitally interlinked 'design creation'.

Following a new design method, projects lead by a common subject versus a trigger point (in this case study in London regarding the term 'connecting' versus sustainable material), meant experimenting with material on the one hand and the knowledge of design heritage on the other hand.

Based on chapter 4.1., the long-term aim to discuss and outline the designers 'driving range' in Europe's future (see the next chapter written by two 'industrial designers' from Great Britain and Germany), there are two specific means for gaining experiences: On the one hand, we could take a conventional education model in which all students are given a common subject for the semester and asked to solve design problems, through defined exercises and design tasks. Further modifications could mean that students learn together with alumni and experts on workshop-based projects that are cross-cultural and involve diversity management. Each project would begin with a kick-off meeting, in open-minded inspirational and creative space with a set framework – yet allowing for new experiences and new inner workings.

On the other hand, a new approach to learning lessons could be implemented – for lifelong learning to serve as the best resource for society. This could be realised by means of **'material-based design thinking' (Wachs, M.-E.)**, as was the content of the workshop realized by Wachs in 2019 at Royal College of Art, London, with a mixed group of students from different educational levels (from bachelor to PhD in design).

Here follows a description of the didactic and content process relevant parameters of this field of study – **'materialising immateriality'** – could be transferred into a creative and educational model, generating the next connected learning landscape for the forerunners in design engineering and other disciplines [48]:

1. **Select a group of individuals from different** education levels who have a variety of interests, originating from various design disciplines and thesis topics. This will **provide a pan-European perspective** with typical views unique to each nation.

[48] Please note: to strengthen the project method, a modified workshop had been planned for VIA University Denmark in September 2020 and was postponed to the end of 2021, because of the global Covid-19 pandemic.

2. **A trigger point is chosen**, such as; a sustainable textile material (in the field study mentioned: a wool-mix cellulose sandwich, paper cord and innovative spacer fabric based on natural materials); media to play with, as a mediating factor; the definition of keywords determined by the group itself (see case study below).

3. As the workshop was formerly named 'fashioning furniture future', the **trigger point for the national** and **culturally** characteristic point of **view** was assigned by the teacher with the material selection, and a design object – in the field study, 'chairs', with examples from European designers and representative objects.

4. **Keyword collection or mind-mapping is done by the group**, resulting from a discussion about **a common subject relevant to society** – in this field study the common subject 'to connect' was defined for the workshop by the group at RCA London in 2019, who's members represented diverse design disciplines and educational levels.

5. **A first interim presentation of the new partnerships is held in pairs or small groups.**

6. **Typical, culturally based 'product language'** is drawn from different national design languages (in this case, chairs from German, Danish, and British Designers were presented and discussed), participants are immersed into the subject of finding a sensible way to integrate national design heritage into modern designs and discuss process parameters.

7. **Individual experiments and design creations are based on background information: Experiments are performed with** the help of the common subject keyword collection, **materials**, the historical background and design object information and key questions.

8. **Experiments are done and there is one-on-one coaching by the teacher (and other experts) for each individual designer, to help them 'materialise' the design concept,** sketch and build models, as well as discussed by sketches, and report about the model for **'materialising immateriality'** ('thinking through sketching with material' or 'visualizing via tangible sketching with the help of fabrics')

9. **A second individual presentation of design concept and products is given,** with a special focus on each individual design task relating to the PhD thesis, or the semester task in the BA or MA programmes.

10. **A first evaluation is made and one-on-one feedback is given at the end of the workshop, addressing the following points: 1) How to move forward with own design / concepts; 2) To reflect on the workshop, new experiences, transfer possibilities, social awareness, awareness of others' perspectives and societal needs in time, relating to culture, taking a closer look at 'other' perspectives** (note by the author: empathy and respect for integrating 'otherness')

11. **Second evaluation and feedback via digital media after two days by both parties: teacher and talents/students.** (see graphic 05: design didactic approach related to the evaluation, p. 157)

In September 2019, the group at the Royal College of Art London, was guided by the German perspective, as well as a German teacher – an industrial designer and design theorist, with a wide range of experiences in various design disciplines. This provided inspiration to the diverse group of women and men (BA to PhD candidates) who represented various design disciplines and different countries (in this case: Great Britain, France, China, and Iran. None of the participants knew each other before the workshop).

The workshop's development and both rounds of evaluation, revealed that the design groups' participants and talents were interested in, and focussed on the following:

— creating design concepts
— studying design – particularly at RCA London
— meeting new creatives with other perspectives and experiences, and different cultural backgrounds
— gender balance of the groups' participants
— all participants showed an enormous interest in new experiences within a design group, using innovative material
— open-mindedness for collaborative and cooperative face-to-face design activities and design thinking
— willingness to present and have open-minded discussions in a design relevant community
— respecting each other's opinion, without regard to previous education and design discipline, without regard to gender or culture
— defining a common goal, developing a sense of design creation, thinking collaboratively (in the case study 'to connect' a very social and culturally relevant design space was created).

At the beginning of the workshop, it quickly became evident that everybody was interested in clearly defining and refining her or his 'own' current design project, in order to draw greater benefit from the workshop, rather than just participating and trying to transferring the experiences – made together – later on.

As such, by participating in this opportunity and taking on the challenge of using a sustainable textile material and joining discussions about 'connectivity', every designer was highly motivated to develop their own special language and create a solution that demonstrated a strong passion for their individually selected haptic material. **The joy of hands-on designing with real fibres, textiles and surfaces, highly benefitted the produced results.**

Please look at these elected results that represent four different design perspectives, as related to the previous design education of each party [49]:

1. BA candidate: designed a chair. (without photo)
2. PhD talent H: with the help of design thinking, a playground of 'togetherness' was ideated. (see photo 2.1 and 2.2)
3. PhD talent S generated a scenario for stages with textile materials and light. (see photo 3.1 and 3.2)
4. PhD talent B materialized the image with textile to express feelings. (see photo 4.1 and 4.2)

1. **BA candidate:** Even in a case where the bachelor candidate designed a chair, focussing on object design, yet using connected functions with new smart technology, his solution appears to be more influenced by the overarching perspective of the younger generations, than by his Chinese cultural heritage.
2. **PhD candidate H** designed a playground with objects, aiming to encourage children and adults to connect in a public space. She designed a round swing construction that connected several swings, whereby each swing was interconnected with the neighbouring one. The scale is approximative 1:10.
3. **PhD candidate S** designed a scenario-based model of a stage, integrating materials with differing value in a playful manner. In addition, a classic lighting design language was used to create different atmospheres and metaphoric value for the user – playing with mixed media.
4. **PhD candidate B** applied the most abstract use of material in visualizing the progressive development of the illness dementia (the topic of his thesis). At the same time a mediation between *different 'spaces' and time conditions* was demonstrated by materialising immateriality through multiple perspectives and the feelings of various people: the ill person, their family and friends, as well as health care and medical staff.

2.1

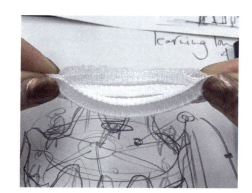

2.2

[49] Please note that H and S and other name abbreviations of those participating in this field study have been used, to respect the privacy data of each student/talent. All pictures shown here: M.-E. Wachs, 2019 – with authorization by the students.

By evaluating the talents immediately after the workshop, it was possible to provide direct feedback. The feedback underlining this pan-European, project-based design engineering education, in the 2020s, focussed on:

- nurturing sensibility and an **awareness of cultural imprints** – cultural education, particularly in design
- obtaining background information about **designing based on prior knowledge**
- **collaborative and cooperative design models**, which focus on analogous human encounters, supported by 3-D stimuli tools and materials in combination with digital design tools
- It makes the process and the **result diffent, when you sketch and communicate your ideas with textile** materials **or** pencils on paper. (see photo 5.1. and 5.2.)
- developing open-minded, **open-ended design solutions, in a highly respectful culturally diverse space, to** integrate everybody regardless of their educational and cultural background – with a focus on designing sustainable and human-centered solutions for a connected era
- a strong **willingness to respect our responsibility** toward nature, diversity, and fellow humans, as well as nurture connected **collaborations for modelling holistic** design solutions
- strengthening empathy and **confidence in design attitudes**.

5.1

3.1

3.2

5.2

4.1

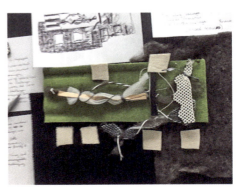
4.2

Two workshops areas took place in 2019: 'TextilePop-Europeans together in sustainable design / engineering / management', in Germany, and 'fashioning furniture future' with 'materialising immateriality' workshop, in Great Britain. In preparation of these workshops, the literature shown in the list was consulted to help fine-tune the design education methods beforehand and provide guidelines for the subsequent evaluations. Upon reflection, **the model of 'materialising immateriality' and the design theory of interdisciplinary practise** could have been implemented more precisely, and will be amended and **applied as follows in the coming years**: a) in different European countries (and beyond); b) in evolving didactic fields of prior education of design engineering, as well as by integrating pupils at schools and children in pre-school; c) by developing immersive, connected digital design spaces.

Thinking by means of materialising, as an innovative design method, is based on materialising immateriality by upgrading the value of hands-on skills. This can be achieved through:

— design education, discourses and experiments in profitable European 'togetherness', i.e. collaborations/partnerships
— revaluing 'sketching' and other hands-on design experiences
— comparing field studies beyond Germany, Great Britain, Sweden and Denmark, and other Northern countries with a similar understanding of design engineering processes
— performing field studies – innovative and creative project spaces in design, materialising ideas through trigger points (i.e. textile and other material-based 3-D samples)
— evaluating common as well as differing design engineering interests
— optimizing digitally connected learning spaces and lesson content: a non-hierarchical cross-cultural range of designers enables open-minded knowledge transfer through design reflection: take a look at design education in the past and especially the associated cultural heritage.

As you could see, the process of generating a 'design creation' – in a didactic sense of connected hands-on and material-based thinking in design, in cross cultural spaces – shapes the future of complex problem solving in a new way. That means, fostering European design understanding could provide even more benefit than we can see today, by transferring the model to other western systems and beyond. However, in order to accomplish this we need more meeting and trigger points in earlier stages of education models – around Europe through analogue + digitally interlinked design creation.

4.3 The European design engineering driving range – Learning lessons for a tangible, non-hierarchical education space towards material and immaterial togetherness (a collaboration between M.-E. Wachs, and A. Hall[50])

4.3.1 Thinking about western design education models – geopolitical changes require new European collaboration for a profitable 'togetherness'.

Keywords
Creative process; European design education; immateriality; future education strategy; non-hierarchical; cross-cultural design space, future experience education strategy, intangible technologies beside tangible experiences.

[50] This chapter is based on a report of EPDE congress in 2019 at University of Strathclyde, Glasgow by Marina-Elena Wachs (Hochschule Niederrhein, Faculty Textile and Clothing Technology, Germany) and Ashley Hall (Royal College of Art – School of Design, London, United Kingdom) together, it was modified by both.

The new 21st century challenges facing design include sustainability, migration, food, water, data security and terrorism. These have left 20th century design approaches lagging behind, while we also now recognise that the 'western' design model is limited. As geopolitical changes accelerate in Europe, new opportunities and challenges are emerging in collaborating for a profitable 'togetherness'. The demand for high-value designed products created across the future European landscape will require new educational talents, working seamlessly across integrated analogue and digital platforms, while responding to evolving cultural needs that emerge as a result of new consumer behaviours. With the help of a differentiated design landscape, we are developing the parameters to meet the future needs of innovation and design engineering opportunities in the 4th industrial revolution: *new standards* in the digitalized learning landscape; *new design methods* for cross cultural creativity and understanding, higher level of integration between qualitative and quantitative approaches in design-engineering, redefining borders and correlations between design, engineering and creativity; knowledge sharing in non-hierarchical cross-cultural learning; differentiating analogue and digital education skill bases in a connected European learning landscape to increase creative diversity. The 'European Designer Driving Range' explores a concept that gives us the possibility to reflect on the needs of tomorrow from a pan-European perspective. We aim to identify the key drivers for a collaborative European non-hierarchical learning landscape and explore how these could be engaged through a future platform.

Much has been written on collaboration and cross-cultural exchanges in product design and design engineering, however comparisons at a European scale, initiated to discuss current and projected future European design issues, are rarer. The 'local' and 'global' are common geographical spaces for design discussions whereas the continental scale is less of a focus.

4.3.2 Comparing national Design Engineering education models – focussing on Germany and Great Britain

Our discussion begins with a comparison between Germany and Great Britain, exploring similarities and differences between design engineering provisions at national levels, to draw out a conversation that focusses on the range of drivers for the future development of pan-European design engineering education.

Let us look back again, to identify common and different characteristics of German and British design education history, to sketch tomorrow's design engineering education space:

4.3.2.1 The Design Engineering model in Germany

In line with the German Bauhaus heritage that focussed on the correlation between art, technique, and the affordance of the social participation of every citizen, design schools like the HfG Ulm – School of Design and other 'Werkkunstschulen' shaped German industrial design education. The parallel awakening of governance, responsibility and the attitude of the Geschwister Scholl Foundation, after WW II, led to the founding of the German Design Council (Rat für Formgebung), in 1953 (see Bürdek, B., 2005). The aim was to communicate the best form and best product language (Steffen, D.) of industrial production, which meant the serious production of ideals during the German Wirtschaftswunder. The mark 'Made in Germany' represented high quality industrial products, with equally high standards in materials and manufacturing techniques, which generated the beginning of long-running profit. Only with designers like Dieter Rams and the enterprise Braun did the German Design heritage in product design – as we know it today – begin to be seen. In other words, German Design is based on the one hand on industrial design (partnering with design engineering), supported by a history of industrial design education that is based on form and function related to Bauhaus artists like Itten, Kandinsky, Albers and Max Bill and their followers. On the other hand, industrial design is also based on craftsmanship at the laboratories of the Werkkunstschulen, in combination with a new consciousness for reflecting design (see chapter 02, e. g. University of Art Braunschweig), with a long tradition of caring about the historical benefits of cultural behaviour. Related to the attention to semantic meaning of products (see Barthes, Haug, et al), after the 1960s, a new driving range of theorists pushed product related messages. The 1980's developed a new spirit of design thinking via historians, like Bürdek, the scientist of design, or van den Boom and Krippendorf, who connected different educational disciplines to unite the design language of products, services and concepts. Besides the universities of applied science, which followed the great pressure of industrial needs, a wider mindset was demanded on a university level, with new terms like the model of 'design thinking', which now bears similarity to the original 'Thinking Design' of Rittel (Rittel, H. W., edited by Reuter, W. D. and Jonas, W., 2013).

The focus went from form, to function, to product language – to storytelling around the world. In Germany this meant design theoretical reflection with the interdisciplinary connectivity between (French) philosophers like Barthes, Foucault, Derrida and architectural theorists like Lampugnani and Zumthor, to name a few. This allowed the cultural and material behaviour towards 'things' to awaken on three levels. 'Things' have to be interpreted according to their meaning as a 3-D object, written object (described by words) and – most importantly today – through images. Today we have an appreciation for products as things and objects, as we view them from a cultural anthropological perspective and look at the relationship construct in terms of its psychological, sociological, historical and cultural relevance, rather than the more technical, functional aspects. By doing so, it provides a holistic view that reveals the meaning of design engineered products and processes, while mirroring the evolution over time. Historical drivers, such as Walter Gropius and students at the Bauhaus 100 years ago, Max Bill and his guest lectures (including Bonsiepe and students at the Hfg Ulm – School of Design), and developments during the 1960's gender and political revolution, illuminated a new view on design education. Only with this new consciousness and the support of powerful institutions and ambassadors (Geschwister Scholl Foundation, Rat für Formgebung, Deutsche Werkbund, to mention only a few) was it possible for society to reawaken a necessary new role for design things with new technological possibilities, new design methods and new aesthetic (life) style. In this circumstance you could point to European sensibility today, and the fact that a 'European Heritage Trust' has been the subject of more serious discussions in Germany, since 2019. Anyhow, the social or political engagement comes into account, when uncertainty is obvious in Europe, but what are the trigger points?

The next great step in design (reflection) occurred in the 1980s, when the new business field of design management became influenced by female concerns, thanks to Brigitte Wolf and a product language that was in accordance with the Offenbacher Manifesto, and which Dagmar Steffen later returned to (see Steffen, D., 2000). **Everything is language!**

This development led to the first doctoral programmes at the end of 1980s in German Universities of Art (e. g. HBK Braunschweig) moving from hard industrial technique to higher cultural value. Again, new perspective, this time with the help of different anthropological approaches for evaluating the next industrial and social and economic revolutionary steps, influencing the curriculum of the universities in design engineering, but in EACH discipline – less by the first ecological movement during the 1970s.

Note
Brigitte Wolf, Alumna of HBK Braunschweig, formulated the first book about 'Design Management', was Professor for Design Theory at University Wuppertal, Germany.

The appearance of innovative formats influenced by the Bologna process led to new university study programmes that focussed on a greater European and international complexity in educational landscapes. However, with the real-time comparability of digital information networking, **a better cultural understanding is needed to work together in digitally connected working spaces wherever we are**. (see graphic 04, p. 148) Everything is language, but language is also driven by a small yet significant group of reflected people, thinkers and doers. On the one hand, design engineering history in Germany is based on high-quality technical knowhow, while on the other hand on very dedicated and passionate people, like the shoe maker Dassler, for example, who built up the Adidas enterprise and later on Puma. The great dynasties of family led businesses from the Lower Rhein region in Germany built up the success of German steel, which gained support from the 'Made in Germany' profile during the 20th century (see Wachs, M.-E., 2008). In addition, former Bauhaus teachers and students who emigrated to Great Britain, after the WW II, served as a connection to the understandable product languages and style in Europe.

Today it is also of great benefit that representative institutions, like the German Society for Design Theory and Research (DGTF, founded in 2003), the Design History Society (DHS, founded in 1977), British Council in Great Britain, and others mentioned above, are supporting an essential view on design. At the same time, the growing cross-cultural publication of articles and books of a more theoretically engaged design community points to good communication among members in Europe.

Reflecting on the common themes for future learning landscapes in design engineering in Europe we realise that the first industrial revolution is what made high-quality engineering possible both in the United Kingdom and Germany. Every time cultural and material behaviour impacts societal behaviour, craftsmanship and engineering education are brought to the next level, with the help of passionate people and the support of councils and organisations. Nevertheless, the terms design and design engineering evolved differently across cultures, though it is not yet clear why. **Germany, for example, cultivates a unique separation between design and engineering**, highlighting how engineering originally related to mechanical engineering and design originally related to creative industries. **This gap will have an enormous impact on design methods within the 4th Industrial Revolution, and future educational landscapes.**

Graphic 04
Non-hierarchical learning landscape for design engineers including domains and partners, graphic: Hall, Wachs, 2019.

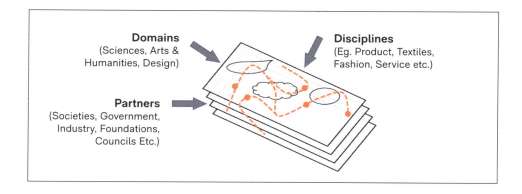

CHAPTER 04

4.3.2.2 Design Engineering models Great Britain

Although the industrial revolution spawned the introduction of art colleges and government run schools of design in the United Kingdom, from the 1830's, the focus was firmly on design 'serving' industrial needs. Design came very much at the end of the process and a latter addition to humanising machines and engineered technology packages. Robin Darwin had proposed bridging the domains of design and engineering as early as 1945, following his role as Secretary to the Training Committee for the Council for Industrial Design (CoID). However, it took until 1980 for the first industrial design engineering postgraduate course to be set up between the Royal College of Art and the Imperial College (see Ewing, 1988; see also Hall & Childs, 2009). Ewing's thesis described the evolution of the design engineering curriculum and pedagogic debate that finally led to the agreement to implement a project-based format, bridging technical excellence and creativity (see Ewing, 1988).

Contemporary developments have seen a move towards a focus on innovation focus, in some engineering design programmes, where design and engineering have dissolved into a landscape of fluid methods that support both disruptive and experimental design-led innovations. While this has brought new opportunities, issues remain and some new ones are emerging for design education.

Increasing technological complexity poses a challenge for the traditional engineering-based teaching of technology principles. For example, technologies that are 'known', versus black box technologies, which need different epistemological approaches based on the comparison of inputs and outputs, and knowledge of restraints. Artificial intelligence is one such field, where it is widely recognised that its 'black box' nature is challenging traditional forms of technology development and exposing new types of risks (see Devlin, 2017). A welcome development has been the increase in cultural diversity and disciplinary crossover from other fields into engineering design, especially at the postgraduate level. This has brought with it a greater variety of creative design methods, new approaches and insights, alongside applied digital and analogue creation and craft skills enriching the field. The challenges that have arisen require a shift from delivering education from undergraduate to postgraduate within the same discipline to a level that allows for wider disciplinary perspectives and a certain number of founding principles, to bring student cohorts up to speed quickly.

This has brought into question the value of traditional design skills and their ongoing relevance, versus cross disciplinary design-led innovation skills. When some students graduate from postgraduate engineering design degrees without the ability to draw, yet are able to express themselves creatively through code, this challenges the longstanding foundations of design education. As technologies like the aforementioned AI – alongside biotechnology and nanotechnology – become more intangible as a result of the material scale of operation and diffused impacts, they challenge long-held traditional 'skill set' models, and **more importantly the idea of thinking through making (see Sennett, 2008). Immateriality and intangibility, alongside new languages for creativity, challenge structures, assumptions, teaching models and the perspective of educators**. Cultural behaviour in design engineering has also emerged as an issue, both from within the discipline in terms of its global fitness for purpose and compatibility with other philosophies (see Brezing, Childs, P.R.N., et al. 2011), and also its tendency for colonisation of creative methods (see Diethelm, 2006), potentially limiting global design variety (see Hall, 2017). In terms of Great Britain's design education, one could argue that the challenge there is that it sees its practices primarily operating on either a local or global scale – largely ignoring European ties.

In our comparison we aim to explore these similarities and differences, to evoke a discussion about **what will drive the range of future requirements for European engineering design education, in the future**. Using different design languages in a variety of material and immaterial learning landscapes in the future requires a common understanding of historically based design behaviours that can be linked to future innovative design engineering challenges.

4.3.3 Case studies – innovative creative project spaces in design – towards 'togetherness'

A number of experimental educational projects and initiatives were developed by both authors of this chapter. In special Hall experienced in the UK in order to tackle some of the perceived issues from teams of cross-cultural multi-disciplinary design engineering postgraduate students. These concerned issues in the areas of developing communications in multi-disciplinary groups, the problem of resistance to failure in design experimentation and developing competence in tangible aesthetic design language. In interdisciplinarity the assumption made in the literature is that we should strive for clear communications and that any errors in communicating ideas can reduce creativity thereby producing weaker design solutions (see Torrisi & Hall, 2013).

Research on missing miscommunications proved that miscommunications, especially at the early creative stage of interdisciplinary cross-cultural design project can lead to new creative capital and bring forward new and impactful innovation concepts (see Torrisi & Hall, 2013). We used a process of c-sketching and analysed group emotional journeys compared to creative outputs. These were then cross referenced across the entire cohort (40+) in order to discover that misunderstanding leads to differences, many of which were not intended by the initiators but later on became new unintended innovation routes.

Failure is frequently avoided by design students and it seems that this is especially the case with those from technical and scientific backgrounds. Often this is for good reason for example in reducing risks in critical structures. **In design led innovation and especially in design engineering failure is a key ingredient on the road to success.** 'I have not failed, I found 10.000 ways that won't work' stated Thomas Edison and the classic Beckett quote 'Ever tried. Ever failed. No matter. Try again Fail again. Fail better.' We developed the Elastic Octopus module (see Hall, Bahk, Gordon, Wright, 2016) where students would succeed in direct relation to their ability to fail and primed the projects with a series of unsolvable experimental aims than asked groups to map their experimental failures. The groups completed a significant number of 20-30 experiments each with a 2-week period. When we reviewed the designs, it was clear that some groups had in fact succeeded yet described their project as a failure by invoking a kind of cognitive dissonance (see Festinger, 1957). Ultimately the key insight came from one student interviewed after the module who claimed that the project improved her creative resilience to failures and that she was much happier to take on future design challenges when the end result was unclear.

The third pedagogic experiment was driven by noting the lack of tangible aesthetic design skills needed for discussing design innovations. The ubiquity of screen based creative tools and the high level 'finished' quality of digitally content has had an impact on tangible form creation by students. We developed a series of tangible aesthetics workshops (see Hall, Ferrarello, 2018) which sought to develop skills for developing analogue analysis for differentiating objective and subjective qualities leading to identifying affordances and signifiers. Although we found that differentiating between objective and subjective features was more difficult for students, they were able to begin appreciating the value of developing their own competence in form language. All three projects taken together indicate that although there are many advantages and positive outcomes for teaching cross cultural interdisciplinary groups that there are also new challenges that need resolving.

The German point of view characterises the innovative learning landscapes of the last ten years of interdisciplinary projects connecting different study programmes – by the author Wachs (design engineering, textile and clothing management, textile technology, product developer, etc.). These projects connected participants from different levels, including BA and MA students, over the course of one semester/year. The experiences gained as a result of the innovative didactic impact of the study programmes were enhanced through the interdisciplinary outlook, the benefits of learning through participation, and a non-hierarchical understanding of how to communicate ideas with different media (see Wachs, 2018). This approach was clearly an effective way to research the subject of smart and sustainable solutions, communicating and mediating in a reflective environment from multiple perspectives, and in a multidimensional concept of design thinking – from a German perspective.

Once a learning landscape, consisting of reflective interdisciplinary and non-hierarchical structures with students within a university educational space had been created, the next step was to develop a non-hierarchical designing and thinking space for students, teachers and external experts to get creative. Secondly it promoted an open, interconnected and cross-cultural analogue and digital communication opportunity on a European level.

In 2019, the cross-cultural multidisciplinary workshops held in Germany (see chapter 4.1.) and Great Britain (see chapter 4.2.), **on the common subject of smart sustainable solutions, allowed a wide variety of experiences to come together, via a new designer driving range in a tangible non-hierarchical learning landscape that took place 'outside the university box'.** This first designing landscape, **created in Europe by designers from Great Britain, Sweden, Switzerland and Germany, precisely underlines the kind of cross-European landscape that needs more support.**

4.3.4 Defining the educational frame for future design engineering landscape strategies

Defining the needs of a future European design landscape, in many ways, bucks the trends of globalisation and localisation by considering a middle scale. The question of what constitutes as being 'European' from a design perspective can be seen from geographical, political, cultural, historical or economic perspective. Recent global geo-political developments, **shifts in power relations, trade imbalances and a move towards digital conflict have reframed the need to understand valuable connections and alliances on a European level**.

To foster a better-connected European design engineering environment, we need to improve the digital university landscape and support role models of new and merging forms of innovation in design engineering.

We don't want to sacrifice the analogue (tangible) skills of drawing and sketching in order to obtain the digital, yet we need to make space in our curricula for this. Drawing is the basis of discussions in many design sectors, as much as coding is in AI, and other technical fields where design has yet to fully engage. **The benefit could be, to connect undergraduates and research students in independent working spaces across Europe with the support of well-connected digital tools.** The aim should be to build European strength for transmitting and **discussing design issues simultaneously, by working together to conceive innovative European smart** solutions across generations – the aim being to create common spaces for designing together – both material and immaterial.

In order to realise this, do we need a European-wide common understanding of designing and design engineering, which looks at the new ethical and methodical challenges of working with artificial intelligence, robotics and the generative automation of engineering practices? According to Schwab and other persons underlined the beneficial factors and economical power of human capital, **we have to invest in a stronger 'togetherness' in our shifted geopolitical circumstances**. Ergo, smart micro and macro factories need **thinkers and talents from all educational levels and researchers, who work together at a cross-European level**.

4.3.5 A non-hierarchical cross-cultural driving range for designers

Projecting future design issues across Europe requires an examination of our design engineering landscapes. There are no longer any universal answers in design and the global decolonisation of design has begun. While there is ample debate on national design policies and the effects of localisation and globalisation, less attention is paid to discussions on a European level, regarding the barriers, challenges and opportunities of future design engineering education. Although European initiatives like Horizon 2020 and Erasmus have been major contributors to knowledge gathering and improving the competitiveness of the EU economy, there has been relatively little focus on the future strategic needs of a specifically European design future and the key drivers that support the needs and issues that must be overcome. **Drawing together conclusions from our discussion on the future drivers for success in training the new generation of European engineering designers, several issues have emerged:** the cultural understanding of different cultural habits and engineering histories; tensions between traditional 'classic' industrial design teaching methods and the needs of new emerging design disciplines, intangible technologies and designing beyond discipline; the need to face interactive / reflective pathways together; financial support from European parliaments for design / engineering / management – new design foundations (e.g. for design engineering heritage); new key drivers with a holistic view on design engineering spaces and education.

The disparity between the national integration of design and engineering disciplines across Europe remains a barrier, and there is also still a need to develop new continental cultural models of design that recognise Europe as a powerful and essential design culture in its own right.

The opportunities for a new future European design landscape include recognising the need to develop a stronger concept of European culture, in terms of design methods that respect and embrace cultural diversity. Furthermore, **our comparable views on design education history in Great Britain and Germany leads us to the next question: Could the concept of a European designer 'driving range' bridge the gap between globalisation and localisation on a European scale?** And, could this offer a new perspective for rethinking the relationships between traditional analogue and the contemporary digital skillsets of the future, with European designers and students working together in a more collaborative digital environment, to create new partnerships?

The benefits of digitalisation in connected study programmes that promote cultural understanding, optimized processes and sustainability, simply cannot be ignored. Our collective retrospection on European design history and industry, as it related to the development of educational programmes, leads us to sketch out a European 'driving range' as an innovative landscape for a tangible, non-hierarchical learning platform that will benefit a material-immaterial design togetherness.

Therefore, we propose focussing on a fundamental design engineering shift in education programmes for Europe.

PRELIMINARY APPRAISAL 02

(by Wachs, M.-E.)

The best Design Engineering models for Europe's future – providing resilient economic and educational benefits by 'forming' fine arts and making the future tangible

When considering the last ten years of European design engineering education – in particular in northern European countries (as outlined in chapter 3 and 4) it is clear that we need to use the knowledge gained from field studies and didactic models, to achieve a holistic education. The field studies revealed the positive impact on students' skills when they were given the opportunity to work in a uniquely conceived environment that was non-hierarchical and interdisciplinary; placed a strong interest on design and respect for 'otherness' (Wachs, M.-E., 2020; based on Paul Ricœur); covered cross cultural themes; promoted gender equilibrium; and exposed participants to cultural design heritage. These students – the upcoming talents – are our capital and will become our future experts. They will be the ones to represent and generate an ongoing and open-minded model for lifelong learning and creation in design engineering disciplines.

We must also consider the results and evaluation of the experiences gained through the design education models outlined in chapter 3, 4.1 and 4.2, which show a very positive development: an optimized democratic framework for teaching design engineering and design theory in interdisciplinary teams and practical workshops. The results were a fantastic range of design solutions, generated on a small scale and in very short workshops. These products were developed in optimized 'learning landscapes' (see chaper 4.1. and 4.2.) that combined freedom, willingness, open-mindedness, low stress (because no direct credits could be earned), intrinsic values and design talent.

M.-E. Wachs's, experience, over the course of more than 20 years of interdisciplinary project-based design theory at universities helped her identify the most fundamental factors for design innovation. Subsequently, these extra-curricular projects underlined the significance of this distinctive and relaxed design studio atmosphere, which enabled the 'flow' in designing (see chapter 6.1). Reaching a state of design flow requires optimized pre-conditions: support for interdisciplinary creativity; a holistic perspective; empathy-based education. This is the most effective method for achieving optimal results that are economically resilient and ecologically meaningful for society.

Besides these optimized design conditions, the field studies performed over the last years in Germany, as well in the United Kingdom and Scandinavian countries, **represent one aspect of the benefits of resilience. In addition, the studies revealed two essential parameters for creating landscape models:**

Firstly, **there is an ongoing trend towards theory-based design and design thinking, based on practiced models. This brings with it the risk of losing important handcraft skills and knowledge rooted in various places throughout Europe, which drive creative design and the creative industries. Yet, due to the digitalization, emotional and manufacturing intelligence are more important now than ever before**, in every aspect of business as well as the increasing need for competence in using artificial intelligence. This – as an example of integrated cultural intelligence – will be discussed further on in chapter 6 (see details in 6.1 and 6.3.).

Secondly, the results each team produces directly, correlates to the individuals' prior education level, particularly in design or design engineering. **All prior knowledge that can be conveyed at schools within the next few years will determine the design engineering competence and success** (in this holistic perspective) of any nation or culture on this continent – or perhaps even the planet. **If we are interested in empowering Europe's design engineering communities to reach a strong capacity and efficiency, we have to invest in the cultural education at school – in preschool we have to invest in forming by fine arts + STEM courses forward to a holistic education**. This will be discussed in more detail – including a 'position paper' - in chapter 7 (see 7.1 and 7.2).

Aside from the 'designers paradise', with optimised collaboration conditions as described above, a return to the cultural heritage of design education is required. We also have **to focus on the prosperity of the futuristic Ps – people, planet, profession, positioning** – and define this prosperity's sources:

1. **Design education as a role model versus its ongoing relation to society**
When art and technology joined forces 100 years ago, it was based on technical generated design products for everybody to use, but with a focus on economic benefits. This is changing now and design education should serve as the foundation.

2. **Pre-school playgrounds as a cradle for design engineer skills'**
Coming back to human's roots and honouring the potential of 'fine arts', in combining new didactical formats of 'manmade' courses, give space to shape and to manage future tangible interconnecting design areas – a whishful thinking?

3. **Design schools are a source of entrepreneurial spirit and influencers.**

The influence of an individual who represents a role model – as a designer or a design engineer and design 'educator'– who also proclaims new parameters, is highly relevant.

A 'person of the hour' can be very convincing and influential with regard to the education and design system. In addition, they can facilitate exchanges with other relevant communities in European countries. With this in mind, moving forward will entail the entrepreneurial power of design schools and designers: around the globe, across Europe, of all genders. After all, they have already demonstrated an enormous cooperative power to optimize models as drivers of design engineering. This is reflected upon and discussed in the following chapter (5), which highlights the female designers who operated 'under the public radar' during the Bauhaus era, for example. The best design engineering models for European's future are giving new value to art and design and its influence on children's skills, and also providing new models for mentoring programmes (see the initiative by Marina-Elena Wachs with the PEM – 'Programme of European Mentoring' since 2021) in design. These are elements of social entrepreneurship that are essential for sketching a resilient future in design (and educational programmes), which we will focus on in chapter 5.

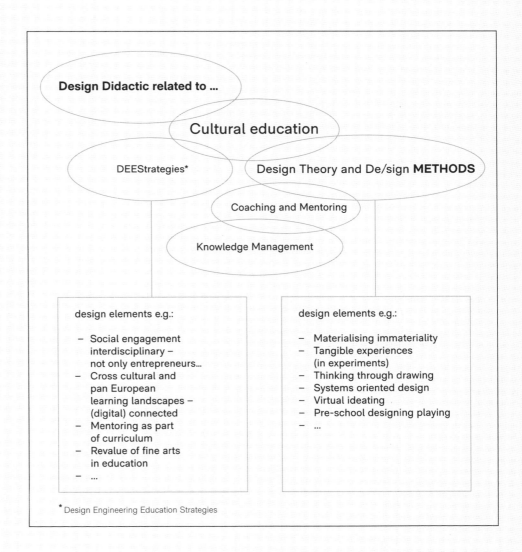

Graphic 05
The Design Didactic Approach, M.-E. Wachs, 2021

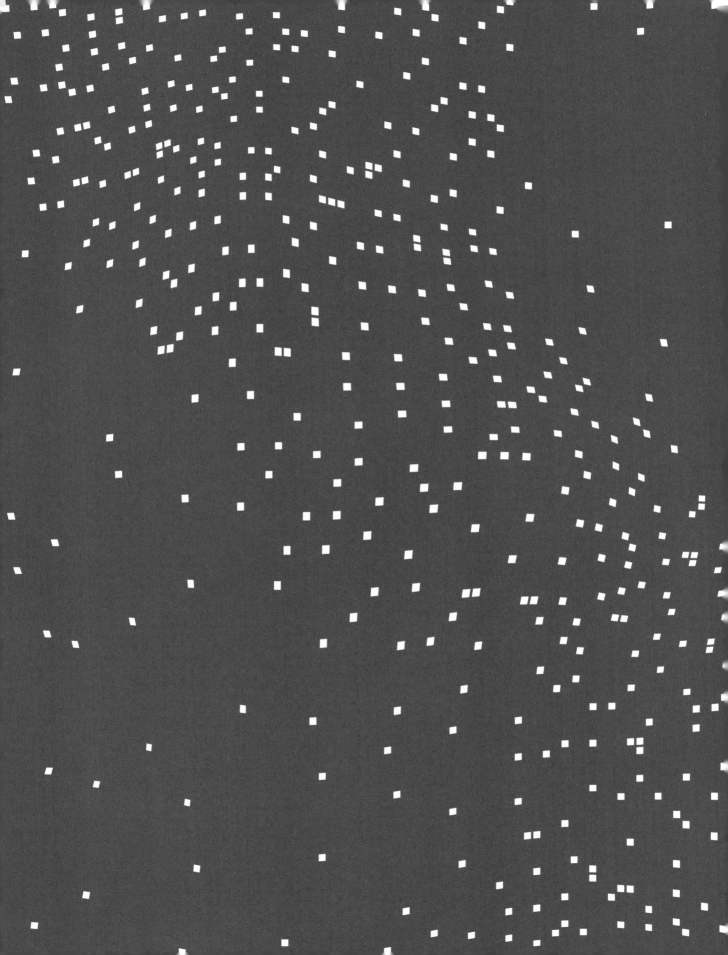

CHAPTER 05

158 — 173
Business models in design
education and their entrepreneurial
power – yesterday's future

5.1 The entrepreneurial power of design schools – the past and future drivers of industry business models

5.1.1 The Bauhaus's influence on industrial design history – the power of businessmen and the female factor

Keywords
European design education and history, the Bauhaus and women's entrepreneurship in architecture and textile, Bauhaus business model partnerships, design education strategies, design schools as business models, researchers' responsibilities and intellectual property, design schools – like the Bauhaus – as industrial design catalysts, future entrepreneurial design schools as business models with the help of foundations.

If we look at the Bauhaus as a business excellence model, we have to look at the power of businessmen like Walter Gropius and Mies van der Rohe, and their influence on industrial design and architecture in practice AND in design education in history. Focussing on the Bauhaus's presence and education system in creative industries, these personalities used their income and their market awareness (see Barcelona Pavilion) to organise the design school. Their buildings demonstrated that Bauhaus teachers were concerned with constructing architecture for eternity – matching the spirit of modernity, but always supported by rich entrepreneurs and an intellectual playground. Teachers at the Bauhaus were employed to create products suitable for the market, but very often profit was actually a result of the 'partnerships' these men had with women at the Bauhaus (see for example Marcel Breuer and Gunta Stölzl cooperation chair weaving), who provided great impulses for innovative designs and architecture. These beneficial partnerships were based on mixed teams, but women were not yet equitable enough to be announced as *official* project partners. Besides this, beneficial economic and practice-based projects mirrored the drive for combined interdisciplinary courses of the Bauhaus to suit a changing market.

Following in the footsteps of these design education strategies, today, we use methods like *design thinking* in interdisciplinary AND non-hierarchical AND cross-cultural creative playgrounds. In succession to the educational heritage of the Bauhaus, we have to develop advanced business models in design and design education.

Remembering the 'dual personalities' of architects and designers like Walter Gropius (e.g. see Meisterhaus Dessau, 1926 and 2018 Bauhaus Dessau), Mies van der Rohe (e.g. Barcelona Pavilion 1927 and 2017), and Hannes Meyer, who all functioned as practitioners + teachers, artists + partners of female students. Later on, they practiced together with female designers and architects as working partners – although these women did not get to sign their work and their contribution was not usually publicly acknowledged in! This phenomenon illustrates the distribution of (gender) power in the creative industries.

It should be noted that a classical education was still standard for women born at the end of 19th century and the beginning 20th century, such as Lilly Reich (1885–1947), Anni Albers (born Fleischmann 1899–1994) (see for example Anni Albers weaving all in', in: catalogue Kunstsammlung Dusseldorf und Tate Etc magazine, both 2018), and other design students of the Bauhaus era. It was a time of classical artistic education at Weimar's Grand-Ducal Saxon Art School (Herzögliche Kunsthochschule), combined with textile craftsmanship knowledge (see Droste, M., in: Pepchinski, M., 2017, p. 105.), when women aimed to marry and run a household. They did not aspire to run their own business in applied art.

It is impossible to imagine the architectural languages of the Bauhaus era without the influence of textiles, tactility, geometric construction elements, which are the **dominant** parts of the **females'** key **signature on interior designs and collaborative architectural projects of that time.**

5.1.2 Relations between creative male geniuses and their ideal female partners – case studies of Lilly Reich, Lotte Stam-Beese, Wera Meyer-Waldeck, Anni Albers

Picture 04
Prof. Dr. Wachs' talks at annual conference of Design History Society (DHS) in Newcastle with the conference theme: 'The Cost of Design – Bauhaus and Business': 'The Entrepreneurial Power Of Design Schools – Like Bauhaus – As Driver For Business Models For Industry Yesterday And Tomorrow – Case Studies: Profitable Partnerships In Architecture And Textile', 06.09.2019, Northumbria University, photo: M.-E. Wachs, 2019.

We have to take an in-depth look at four *genius* partnerships, which were examined in case studies, in order to understand the retrospection and the various ways in which the entrepreneurs profited – as did society and the women who were behind the scenes and 'under the public radar'. Last, but not least, these studies show how these women's relationships paved the way for **female run businesses in interior design and architecture today.**

Lilly Reich's (1885–1947) working collaboration and romantic partnership with Mies van der Rohe is the best-known case study, as demonstrated by the **Café 'Samt & Seide', in 1927 in Berlin** (see portraits of both and the café, various pictures in the archive of Lange, Christiane 2013, ibid: Droste M., in: Pepchinski, M., 2017; ibid: Oswald, P., Bauhaus-Zeitschrift der Stiftung Bauhaus Dessau, 2014), seen in parallel to the fair 'Die Mode der Dame'. It is a well-known fact, that the **open-plan spatial arrangements** they both worked on together until Mies van der Rohe emigrated to USA, like the design of the Café, demonstrated their understanding of business strategy. They were a couple of enthusiastic creators and designers, who drew their inspiration for their everyday work from art. Yet, they also relied on various individuals and institutions: Mies van der Rohe was affiliated with entrepreneurs like Hermann Lange, who was the owner of a large successful silk production enterprise in Krefeld, Germany.[51] Lange gave van der Rohe different opportunities to work on architectural projects in the local scene in Germany, and beyond.

Simultaneously, Lilly Reich's influence (see portrait and glass expo at Stuttgart, in: Pepchinski, M., 2017), which resulted from their representative work for the Deutsche Werkbund, the Frankfurt Fair and others, had an enormous impact on different creative scenes at that time. This included, but was not limited to, flexible mobile textiles for interior architecture. The exhibitions showcased new models of *thinking spaces*, which were demonstrated very well and were underlined by the designs of Anni Albers (see following sections).

Lotte Stam-Beese (1903–1988) was a student of the Bauhaus and in a relationship with Hannes Meyer, one of the directors of the Bauhaus at that time. She had been married to the architect and designer Marten Stam, from whom she later separated. These unique partnerships, evident in several biographies, illustrate the way in which female architects reached their business potential in former times – at the side of a powerful man, be it a designer or an architect. *'In 1928 Lotte Beese was the first female student to take part in this new architectural program (by Hannes Meyer). It was not only design and construction that appealed to her, but also the social and humanistic goals Meyer applied to architecture. One of his main principles was 'volksbedarf statt luxusbedarf' (the needs of the people instead of the need of luxury).'*[52] As for other women in the construction business, social housing became a main task, but no woman had ever reached the position of an urban planning architect in Rotterdam before. **With her view as a German architect after World War II, she was able to express her outstanding actions and way of thinking – a tribute** to her career in the creative applied fields. It was her human perspective on how to integrate social classes within the framework of city life that sparked her incredible cross-cultural cooperation, resulting in a light-infused space for human beings to share. A space for people as 'residents' with a consciousness of shared 'land' and 'mothers earth' – shared gardens and grounds that are everyone's responsibility. This was a very feminine view at that time – in architecture and urban planning.

Wera Meyer-Waldeck (1906–1964) received her Bauhaus diploma in architecture in 1931. Having studied at the State Academy for Applied Arts in Dresden for three years, she then studied at the Bauhaus for another four years, with the aspiration to learn the Bauhaus aesthetic: long lasting, classical design translated into modernity for eternity.

During that time, she also worked at Walter Gropius's and Hannes Meyer's architectural offices. *'She was among the few people to experience all three of the Bauhaus' directors. Her projects were of an outstanding quality, and during her student years she was able to collaborate on the interior planning and furniture designs for the Office in Dessau by Walter Gropius and the German Trade Union Confederation School in Bernau by Hannes Meyer.'*[53] Wera Meyer-Waldeck was inspired by female architects at the Massachusetts Institute of Technology (MIT) in Boston,

---

[51] See Lange, C., 2013, in: Brüderlin, Markus (Hg.), 2013, Kunst & Textil – Stoff als Material und Idee in der Moderne von Klimt bis heute, Kunstmuseum Wolfsburg, Hatje Cantz Verlag, catalogue to the same named exhibition 12.10.2013 - 02.03.2014, p. 134.; ibid: Droste, M., 2017, The Creative Pair: Lilly Reich and the Collaboration with Mies van der Rohe.

[52] Oosterhof, H., 2017, Lotte Stam-Beese. A Committed Architect and Urban Planner, in: Pepchinski, M. et al, Deutsches Architektur Museum, 2017, Frau Architekt, p. 180.

[53] Hervas Y Hervas, Josenia, 2017, A Bauhaus Architect in West Germany: Wera Meyer-Waldeck, in: Pepchinski, M. et al, 2017, Frau Architekt, p. 167.

USA, concerning the integration of solar energy into building concepts. **In 1955 she built Dr. Bockemühl's Residence at Beuel-Limperich** (see Meyer – Waldeck Bockemühle Residence AND solar house USA, in: Pepchinski, M. 2017). **This underlines how private individuals enabled females to gain power** in architecture, by accepting new business models, as well as very alternative female role models in society. Aside from the revolutionary and outstanding collaborative power she achieved with respect to her teacher and directors, she also fought passionately to work as an architect – **not** as an interior designer. '[…], *as men increasingly were conscripted to service on the front lines, she was able to take on more responsible employment, including work on bridge construction (the Elbe High Bridge Hamburg), construction supervision for the German National Railway […]*'[54] and other projects, as examined by Hérvas Y Heras in 2017.

Anni Albers (born as Anni Fleischmann, 1899–1994) could be considered as a visionary role model, embodying the secret of business success: the connection between (textile) artist and designer.

Both the support she received from Josef Albers – her former teacher at the Bauhaus and later on her husband – as well as their highly productive and inspirational exchanges and collaboration as partners in the USA, are what served as the catalyst for her becoming the first female designer with a one-person exhibition at the Museum of Modern Art in New York, in 1949. The art historian, curator and editor Simon Grant stated that she was '*Not only a consummate teacher of her discipline, she also revitalised the ancient craft, creating beautiful works of art to rival any of those made by her contemporaries.*'[55]

In the context of that period, both in the USA **and** Europe, it must be acknowledged how great an influence the female factor had on business success. Nevertheless, being the best in their field as a woman was not enough for them to receive recognition. It was still essential for them to have the support of powerful men – architects, artists and designers.

Like Lilly Reich, Anni Albers designed many different types of room dividers throughout her career, applying them **as integral architectural elements. She defined them as 'pliable planes'** (see wooden paravent hanging, exhibition Anni Albers 2018, in Dusseldorf) that always incorporated **craft-like qualities (i. e. weaving or textile constructions) into the applied arts, while simultaneously experimenting with the interplay of material and light**. Even her diploma work in the weaving laboratories at the Bauhaus used unconventional materials like strips of cellophane (see reconstruction of A. Albers Diploma at the exhibition Dusseldorf 2018). She used these threads to reflect light and create a shimmering effect, yet at the same time to build a very energy-efficient wall or window drapery thus achieving a high-quality interior design. '*The advantages of woven walls are that they are lightweight, mobile, and adaptable.*'[56] Curator Angela Wenzel, from Kunstsammlung Nordrhein Westfalen, drew parallels between the work of Anni Albers and Gottfried Semper (19th century), and equated walls made of textile materials with **woven materials as the mother of architecture**. Wenzel's idea is supported in Anni Albers's essay from 1957, 'The Pliable Plane', which was published in 'Perspecta' – the Yale Architectural Journal, in which Anni Albers called for a collaboration

---

[54] Hervas Y Hérvas, J., 2017, A Bauhaus Architect in West Germany: Wera Meyer-Waldeck, in: Pepchinski, Mary et al, Deutsches Architektur Museum, 2017, Frau Architekt, p. 167.

[55] Grant, S., 2018, Tate Etc Issue 44, Tate Gallery, London, editorial, p. 10.

[56] Wenzel, A., in: Kunstsammlung Nordrhein-Westfalen (Hg.), 2018, Exhibition Guide, Anni Albers 9.6.-9.9.2018.

between weavers and architects (see white curtain by A. Albers at exhibition at Dusseldorf, 2018). *'She firmly believed that humanity's first architectural structures had been made using textiles. Tents have existed for more than 10.000 years.'*[57]

Over the years, Albers demonstrated the point of intersection of **'textile art' and 'textile design'** in applied art numerous times in her collaborations with the Knoll enterprise, for example.

The art historian Christiane Lange proclaims the term 'textile Wandsegmente'[58] (Lange, C., in: Brüderlin, M., 2014), translated to 'textile wall segments', to underline the constructive architectural element of textile walls. With the help of **textile wall segments, it was possible to create different spaces, for example, as previously seen in the 'Café Samt & Seide'** in Berlin, in 1927. This was displayed during the exposition 'Die Mode der Dame', which was initiated by the textile entrepreneur Hermann Lange from Krefeld, Germany. He was the head of the Society of German Silk Weavers at that time. Christiane Lange – Hermann Lange's great-granddaughter, has examined the linkage between the Bauhaus and the silk industry from today's perspective: *'In a nutshell: What connects the Bauhaus to the silk industry? Ultimately, it is the common belief in the power of crafts, arts and creativity to facilitate change, even within the very pragmatic context of economic trade.'*[59]

To sum it up, and evaluating the 'relationships of creative male genius and their female partners' we can consider: When textiles become architectural elements, it comes down to the **interplay between their practical functions and characteristics**, e. g. elements that are lightweight, mobile, light reflective, and have the ability to separate spaces through flexible structures while offering acoustic benefits. These are **'representatives' of the textile heritage that form a modern medium through** objects and icons. 'Representatives' refer to both such textile artefacts as well as past entrepreneurs – (creative) male geniuses, like van der Rohe or mentor Hermann Lange, who were influenced by the interplay of ideal partnerships with creative women and the female factor that led them to create success together. **This is how art can be integrated into everyday life. We can reference** the inside of the Barcelona Pavilion 2017 and the project 'Mies Missing Materiality', by Anna & Eugeni Bach, in 2017, where textile art objects demonstrate detailed creative experiments using textile samples within the framework of the exposition. Textiles have a different effect compared to architectural elements for interior design, particularly when they are used as a third skin in the field of construction or on modern-day architectural masterpieces, as demonstrated.

Nowadays, textile is considered an important construction element for architecture, thanks to the women who paved the way for textile representation in 'hard' business fields – formerly reserved for men! Please, make note of this gender aspect, as we still see this in 'hard sciences', associated with STEM courses and considered primarily male, whereas the humanities are primarily thought of as female[60] (see von Braun, C. / Stephan, I., 2009).

---

[57] Wenzel, A., in: Kunstsammlung Nordrhein-Westfalen (Hg.), 2018, Exhibition Guide, Anni Albers 9.6.-9.9.2018, p. 14., ibid: the Josef and Anni Albers Foundation, 1957, the Pliable Plane – Textiles in architecture, online: https://albersfoundation.org/teaching/anni-albers/texts/#tab4

[58] Lange, C., 2013, in: Brüderlin, M. (Hg.), 2013, Kunst & Textil – Stoff als Material und Idee in der Moderne von Klimt bis heute, Kunstmuseum Wolfsburg, Hatje Cantz Verlag, catalogue, 12.10.2013–02.03.2014, p. 134.

[59] See Lange, C., 2019, online 9th of February 2021, https://www.baunetz.de/meldungen/Meldungen-Christiane_Lange_ueber_den_Krefelder_Jubilaeumsbeitrag_6427638.html, translation by Louise Huber-Fennell, 2021.

[60] See Braun, Christina von / Stephan, Inge (ed.), 2009, Gender@Wissen – ein Handbuch der Gender-Theorien, Böhlau, UTB, p. 9 et al.

5.1.3 Retroactive effect of industrial design education brands in history and the future – How educational models benefitted from the Bauhaus legacy and how we can transition to 'education model 4.0'

If we look at our design and educational legacy, we can identify two significant reasons why brands and entrepreneurs should recognize and support the importance of the crafts and their inclusion in academic education: First, to ensure their businesses' future success and generate profit for reinvestment in talents and secondly, to raise awareness of our heritage as an industrial catalyst.

Besides this evaluation, the economically advantageous and practice-based projects reflected the drive behind the Bauhaus's interdisciplinary and collaborative courses, designed to meet the demands of the changing market as well as the need for a change in educational content. First and foremost, the Bauhaus demonstrated an open-minded interplay and interdisciplinary linkage between education levels, by bringing together students, teachers, artists, and experts from architecture and design.

If we look at the Bauhaus's educational model, and that of its successor (the HfG Ulm School of Design), the most significant benefit that has been drawn from this model is the acceptance of the interdisciplinary and holistic view on tasks, which can be applied to art models and the theoretical reflection on design and everyday life solutions.

Today's design education strategies involve using methods like design thinking in interdisciplinary AND non-hierarchical AND cross-cultural creative playgrounds. This follows the path set by the Bauhaus's educational heritage, and is currently leading us to advanced business models in design and design education. We should not 'try' to implement such strategies – that simply means they will not be taught or become commonly accepted as a model. We **must** use them! Just a brief look at the current social challenges and those we will face in the coming years through digitalisation and migration substantiates this.

We need a more interlinked togetherness, which does not just rely on the help of digital devices. We need an understanding of each nation's design history, cultural background and the history of design education! We need to fight, not only for interdisciplinary projects, across different student program levels – from bachelor to master students – who can learn from each other. We also have to continue fighting for non-hierarchical and cross-cultural teams in design: European students together with experts and teachers, artists and children. Future success means listening to and respecting different and diverse ideas on how to create sustainable design solutions. The open-minded female influence of the Bauhaus women – who operated 'under the radar' – was a key component for developing a new role model in design and architecture, as we see today. But in a wider sense, their pioneering work – as well as their unique views on an array of design aspects – also paved the way for non-hierarchical design teams. This was and remains a significant benefit to interlinked societies and design communities – as well to the industry 4.0.

5.1.4 The future value and risk of industrial influences on academic design models and intellectual property

Coming back to the four genius partnerships examined in the case studies can help us understand the mutual benefits both the entrepreneurs and the women behind the scenes enjoyed in various forms, as well as the benefit to society. As mentioned previously, it should also be noted that these collaborations have had a significant and lasting impact on females in interior design and architecture businesses, even though they did not become truly relevant and highly appreciated until 100 years after Bauhaus. **We must recognise the economic, educational and societal benefits that the 'creative industries' offer! These industries and applied arts, triggered by fine art, draw from the knowledge and understanding of craftsmanship and cultural anthropological value.** Does that mean it is time to give up the German (European?) perspective that differentiates between the economy and creative industries? This is the reality we have to mediate and teach the next generations! It is about the interlinking of art expositions and fairs, as well as events which are now curated, like art inspired spaces in fashion (curated concept stores like Andreas Murkudis in Berlin for example).

We have to show appreciation for the cultural value and the benefit of societal education in design and art. Hanno Rauterberg, who holds a PhD in Art History, sets the centenary celebration of the Bauhaus in 2019 in context with the avant-garde De Stijl Movement in Amsterdam, and the Constructivists in Moscow.

They all followed the same model: reduce the design to the most basic construction elements and combine these with playful easiness to achieve modernity.[61]

Reduction and elementary work focus on form, material and object, aiming to create natural structures that promote a societal lifestyle and culture. This meant creating a **'natural order' with the help of elementary geometric forms and pure elementary colours; building upon serial participation to reach uniformity after World War II; creating a new culture and new education system.** This common 'code' of creativity is part of our education legacy.

It began with the Bauhaus's concept, an ideal created and defined by a person like Walter Gropius, who was an architect and warrior for modernity. By enabling male students to work together **with female students and opening the school and typical male-orientated laboratories to women (and not only for the upper-level class), he created a modern educational system, as seen as paradicmatic case?**

Subsequently, architects, designers and entrepreneurs like van der Rohe, Gropius, and Meyer gave women the opportunities to perform hands-on experiments at school and gather experiences in a 'hard' business field. In this manner, they gained insights and an unprecedented awareness of society's needs and the general feelings of that time. This was solely thanks to one person – or his institution – that convinced and defended the **idea** of participation, in order to develop a new culture. However, today we recognise that the success of the men mentioned above depended on women serving as muses and patronesses (see Muscheler, U., 2018, ibid: Müller, U., 2009 et al.).

Aside from the educational benefits, if we look at the creative industries today, it is the dedication of enterprises, like the House of Dior, or entrepreneurs, like Brunello Cucinelli, that are safeguarding the (textile) design heritage. At the beginning of the 21st century, Brunello Cucinelli built up an academic education centre to secure the next textile generation – workers supporting the house brand – thereby incorporating a strong identity into the family-owned enterprise nearby Perugia, Italy. This is only one example that mirrors **the consciousness in the textile industry, and within businesses run by entrepreneurs who were looking for sustainable structures for the future. At the same time, this showcases the human need to invest in the mediation (transfer) of knowledge. This means transferring the value of cultural heritage and craftsmanship** into the best education possible, as well as transferring the knowledge of old and modern methods of industrial production thus generating economic profit.

Cucinelli projected a family atmosphere that was rich in tradition, with the help of their marketing campaigns and images, like those from 2013 (see the enterprise's campaign in 2013 'we are family', dinner table with all members of the enterprise). In the years that followed, the brand developed a more abstract picture-based story, using icons to communicate and representing societal values without focussing on the product. Cucinelli achieved this, for example, by not highlighting the clothes constructed with unique and outstanding cashmere, but by drawing attention to storytelling through the use of historic insignia and pictures from the

---

[61] See Rauterberg, H., 2019, 'Ins Zeitlose entrückt – Das Bauhaus wird hundert, und die Deutschen feiern es geradezu rauschhaft. Noch immer gilt die weltberühmte Kunstschule aus Weimar als fortschrittlich, radikal und widerborstig. Doch stimmt das überhaupt?', in: DIE ZEIT, 17th January 2019, p. 37.

Middle Ages instead. **It is about people's values, and how they identify with a global model and a 'place to remember' – like the ground-breaking Bauhaus model – yet one that is based on a holistic view** (see Brunello Cucinelli promotion campaigns: storytelling – novel from Renaissance time, see ZEIT Magazin, 2016 and others in 2014, 2017 and 2020 in AD and other magazines).

If we look at these past phenomena – the pioneering initiatives and advocacy work of foundations in the fashion and textile industries in Italy and France, who had different interests – we can conclude that the German textile industry has the potential to be at the forefront of the German industry, if they act now. Together, the brands and representatives from the creative and textile industries and institutional testimonials can unite, in the form of museums and other educational clubs (like the Fashion Council Germany, Prince's Foundation GB, German Meisterkreis, German Design Council, and others). Thus, they should begin to build up their brands' own foundations that advocate textiles. In Germany we have such a great community of foundations that it must be possible to work together in a non-hierarchical manner, in an effort to strengthen Europe with respect to sustainability and connected design engineering landscapes (by respecting craftmanship). The f**ollowing collection of parameters identify trigger points that are considered tools and industrial catalysts for the future of (textile) design engineering. Ultimately, the textile system could be transferred to other design engineering businesses, by:** (see graphic 06, p. 169: stakeholders of the future, Wachs, M.-E., 2019)

— saving design history in different forms of media, with key 'knowledge-saving' 3-D models, original prototypes
— transferring craftsmen's knowledge
— mediating in education systems – promoting experts and cultural education for society
— empowering entrepreneurial spirit and enrichment
— constructing value for society and culture in Europe
— developing sustainable solutions and new creative power
— developing advanced economic business models for the post-digital era
— retaining analogue knowledge that is needed to communicate and understand the world model.

So, if we consider the question of intellectual property, we must admit: it is a risk to only look at one person's idea, or **one** institution that is at the forefront, or one entrepreneurial design school. This was made evident in the past with Walter Gropius's model.

However, this should not stop us! After all, most business models are based on one person's idea and passion: The textile academy, created by Brunello Cucinelli, which influenced business models within the industry and trade (even at the stock exchange), is only one such example. Another example of how an entrepreneurial business model can support educational systems in design and applied arts can be found in powerful industrial institutions like the 'Comité Colbert' in France, as well as the growing influence of the Fashion Council Germany (founded in 2015) and the German Meisterkreis (founded in 2011). A new and beneficial foundation, founded in 2020 by Erck Rickmers, is The New Institute in Hamburg, Germany (with sister institutions in Manchester, Venice, California and Berlin), which is taking responsibility for our contemporary cultural problems with a focus on humanities.

These examples can guide us on how to create a new foundation that is essential for saving and securing human knowledge, so that a transition into design innovation – to be produced by the next generation of experts – can be secured. This foundation will also help communicate mile stones in design (engineering) history, much like the writings of the Bauhaus women before.

Who will take the position at the forefront? We need idealists, pioneers to take creative responsibility for the world's problems. Idealists as managers, entrepreneurs, leaders of design schools, industrial drivers and partners with an equally strong sense of duty and sensibility towards creating a good future.

As we have seen, the creative power of all sexes, in combination with different cultures and generations, are the key drivers in design and they influence the entrepreneurial power of business models – in Industry 4.0 and education model (forerunner) of the future. Thus, in order to ensure a brand's sustainable profit, we must create modern workplaces, develop suitable education models and transfer lasting knowledge. In addition, we must raise awareness of our European-based heritage in textile and architectural businesses = in creative industries everywhere. This heritage has an enormously impactful and ethical value for our future society – and it will inspire and encourage new pioneers.

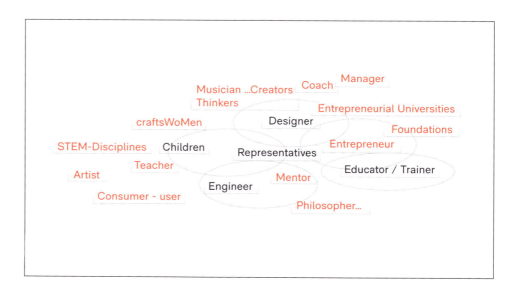

Graphic 06
'Interplay of different stakeholders (interests) – to mention a few...',
M.-E. Wachs, 2021.

5.2 Sustainable thinking in design education AND design engineering industry: the next academic and business models and the 'Management of Knowledge'

Keywords
Sustainable education models; sustainable thinking and acting in design and design education; cultural value; collaborative models in design thinking, design doing; placing trust in new generations of design engineers and design managers; encouragement and motivation as key factors in Management of Education Science's.

Design is sensible and sensitive. Design is the mediator between products and users, and works in the interest of the entrepreneur and the consumer group. In the future, buying behaviour will no longer depend solely on emotional, visual product values. Instead, products will need to embody ethical value, by using resource-efficient production circles, in addition to being functional and long-lasting cultural goods.

High-quality products and services are a consequence of sustainable designing, as it seeks to discovers the origins, needs and additive design qualities, while focussing on responsible behaviour. You have to think, produce, and invest in a sustainable manner in order to act responsibly in production and economic circles. It means developing and implementing a circular thinking model. This is the only way for all participating partners to benefit – including Mother Earth, of course. *'We believe in giving back, whether to Mother Earth or the society that nurtures us.'*[62] This core value, presented by Sonia Sin and Frans Schrofer, expresses our common attitude toward design, as regularly discussed in design circles. A group of students from the Textile and Clothing Technology programme at the Hochschule Niederrhein debated the product language and the material of the 'Infinity' chair (see pictures, p. 173), designed by Frans Schrofer. This prototype was produced together with Fritz Becker GmbH in Brakel, Germany, and was presented at the Interzum fair, first in 2017. The students asked themselves whether 'Infinity' could possibly satisfy long-lasting design qualities in the same way that the ground breaking classic chairs have done in the past.

The case study, which was a co-production between Schrofer Studio and Becker, reveals an enterprise that continues to be a hidden champion of formal bentwood production, and is now also qualified for formal felt production.[63] This proves once again what creative 'co-experiments' can achieve when designing advanced sustainable products and concepts together.

This illustrates that, in the 21st century, we have returned to the complex and valuable qualities of the *'Kraft der Kooperation'*[64] (power of cooperation). This means reaping the benefits of creative joint 'design thinking'[65] or 'thinking design'[66], which was also demonstrated by the insightful and valuable corporation of engineers at the Deutsche Fraunhofer Gesellschaft and design students who worked on and thought about sustainable solutions and design services.

[62] Schrofer, S. and F., http://www.studioschrofer.com, online 1st 07 2017.

[63] See Fritz Becker GmbH Brakel: https://www.becker-brakel.de/en/

[64] Neugebauer, R., Präsident des Fraunhofer Instituts im Gespräch zum Statement 'Design muss Teil der Entwicklung werden', Design Report 2/2017, p. 82 f.

[65] See Method Design Thinking by IDEO, Plattner, H., 2009, Design Thinking, mi-Wirtschaftsverlag-Finanzbuchverlag, München.

[66] See Rittel, H. W., (1992), edited by W. Jonas et al. 2013, Thinking Design, Birkhäuser Verlag, Basel.

The two cooperations highlighted here define the ideal for our society, and for the world, as they were proven to be successful in design history. The German Bauhaus is a comparable institution that utilised this archetype model for creating products, and also boosted trust in the emerging (long lasting) classic design products, at that time.

In today's strange and dynamic times, in which great efforts are being made to 'reinvent the wheel', we should trust our past experiences and appreciate this heritage more powerfully as a real resource. This argument about more 'self-consciousness' is underlined by the futuristic philosophical speech by Roger Willemsen in 2015, when he mentioned, that we came along with a technical mastering and becoming more concurrently, more sovereingn and more powerless, saver and less stable, more target-oriented and diffused.[67] Willemsen stated the consciousness or the feelings of European society, encouraging the younger generations to reject the conventional!

'Sustainability – and nothing less – must be the minimum standard! For the last 15 years, across all faculties of teaching and practical work – including corporate universities in Europe – we can see that people have been passionately focussed on one constant factor: sustainability.' (Wachs, M.-E., 2019, EPDE Conf. Glasgow). And **we trust that our students will increase their understanding and know-how of ecological design, which will lead to new industrial standards and unpretentious living conditions.**

'In order to develop these new standards in ecological design circles, we need to incorporate teachings about natural ways of living in sustainable environments long before university level.' (Wachs, M.-E., 2020, 'FASHION | ZUKUNFT Talks' – e-conference).[68]

Sensible and sensitive sustainable behaviour requires us to support more innovative curricula from the onset of school education, by adding ecological and design engineering to curricula. This will help us remain competitive – and our planet will benefit.

In addition, we face the great paradigm of know-how transfer (in German: 'Wissenstransfer')[69], within the scope of the digital revolution, which Industry 4.0 is demanding. This can be facilitated with the help of valuable transformation design systems (see Jonas, W., 2016). At the same time, it can recover experimental designing methods with multidisciplinary teams of entrepreneurs, engineers, civil actors and designers, who have different cultural backgrounds and are committed to improving society. Besides the positive economic and ecological benefits, this 'Management of Knowledge' has become an increasingly important research topic over the last years. However, now, it is time to also bring the **'Management of Education Sciences' into focus**.

In accordance with John Elkington's 'triple bottom line' (1979) – planet, people, profit – it is time for the younger 'generation app' to get in the running. We must believe in our talents, who will be the leaders of change for sustainable life systems, and the responsibility for sustainable design and sustainable design management will lie in their hands. We must trust that we will profit from sustainability education: the intellectual capital we need in order to save and protect social capital, our existence and natural sources. We must take care of sustainably-educated design

---

[67] See Willemsen, R., 2017, (2016), Wer wir waren – Zukunftsrede, 6. edition, S. Fischer, Frankfurt am Main, p. 44.f, translated by Louise Huber-Fennell, 2021.

[68] Note: the second conference of Prince's Foundation and Fashion Council Germany held in 2020 as e-conference, brought experts and students together in the case of the future of sustainability and -education, see Fashion Council Germany, 2020.

[69] See Präsidium der Hochschule Niederrhein, 2017, Hochschul-Entwicklungsplan – den Wandel gestalten 2021, Hochschule Niederrhein, Präambel, p. 8 f; ibid: Funken, C., 2016, Sheconomy, Bertelsmann.

talents and design managers, by supporting them in their personal and character development. This can be achieved with the help of Gestalt methodology's facilitation of skills and cultural intelligence: **cultural education is the core value of sustainable concepts**. Yet, we must also gain support from the cultural ministries and encourage them to continue attracting innovation-based school curricula. Furthermore, they must be called to redesign pre-education levels – form pre-school onwards – including programmes for Management of Education Sciences and 'Management of Knowledge' as a differentiated research field.

From the mentor's perspective and with understanding as advocate for outstanding talents and cooperation partners, we must all take on responsibility in our everyday lives: a dedication to sustainability in the form of 'skin in the game'[70] (meaning to fully invest) – a concept which will be discussed in chapter 7.

Not taking responsibility for resilient education in industry and at school immediately will have severe consequences for 'sustainable design thinking' in the future! We have to act and collaborate now – while keeping the de-industrialisation, as mentioned above, in mind. This step in the 4th Industrial Revolution is our greatest challenge right now.

---

[70] See Sandberg, S., 2013; ibid: Taleb, 2018.

Picture 05
Sustainable beneficial partnership: Joachim Schelper, (Fritz Becker GmbH & Co. KG: research & development), Frans Schrofer (Studio Schrofer, industrial designer / owner), Dr. Ralf Becker (Fritz Becker GmbH & Co. KG: owner) to present the new 'Infinity Chair' by Frans Schrofer at Interzum fair 2017, Germany. Pic: M.-E. Wachs, 2017.

Picture 06
Guided tour at the booth of enterprise Fritz Becker GmbH & Co. KG, Interzum fair 2017, Germany, pic: M.-E. Wachs. 2017.

Picture 07
Infinity Chair designed by Frans Schrofer, Studio Schrofer Den Haag, pic. M.-E. Wachs, 2017

We thank:
www.studioschrofer.com
www.becker-brakel.de

Pic. 05

Pic. 06

Pic. 07

CHAPTER 06

174 — 207
Our aim to build and use cultural intelligence, using creativity to create ethical and holistical education landscapes

6.1 The 'flow' in design – an extreme super power for working conditions or simply wishful thinking?

6.1.1 The flow in design generates profitable economic and social benefits

Keywords
The flow in design; artificial and artistic intelligence; naive view in design; integrative cultural intelligence; more humanities in design education.

The previous chapter outlined the needs for taking responsibility in sustainable thinking in earlier times. Also, expressing a greater dedication and 'skin in the game' for a resilient pre-school education, will provide the necessary pre-requisites and allow wishful thinking to take place in terms of the next level of design (education) environments. Imagine the 'flow' (Csikzentmihalyi, Mihaly, 1990) as an integral part of designing. Displays of human super power and strength in sports are understood as biological, neuroscientific linkages, performed with ease. This actuality is corroborated by the concept of 'flow' in the field of psychology. This explains how an expert who feels confident of their skills in a particular discipline can tap into exceptional powers, by drawing from their long-term experiences in a moment of (cognitive) relaxation.

Perhaps you have experienced this as designer: getting into a creative flow while in a stress-free environment. In contrast to the way in which Industry 4.0 and the use of artificial intelligence (AI) has developed, the flow is based on emotional intelligence (EQ by Goleman, D.). Hence, this is the hypothesis of this chapter: **investigating emotional and 'artistic' intelligence in cultural education can help define a wish list to improve extreme working conditions, and the flow in design will provide profitable (economic) and social benefits for the post-digital era.**

6.1.2 The flow as a methodological concept in design for complex problem solving

In 2016, the World Economic Forum proclaimed that human skills – the highest of all business skills – will be required for 36% of the 'complex problem solving' tasks (Schwab, K., 2016, p. 42, e. g. 19 % for 'social skills'), according to 'Skills Demand in 2020' and the 'Future of job's report' by the World Economic Forum, as we move into the next decade of digitalisation: Schwab and several authors are using this point to argue that changes in working conditions are inevitable (see Günthner, R., 2019; ibid: Diefenbach, S., 2016). Complex problem solving is not a new phenomenon in design (see Norman, D., 2011, ibid: Simon, H., 1979). However, as people's tasks change – during and after the digitalisation and fourth industrial revolution – the human ability to solve problems will be more relevant than ever. This is due to the fact that artificial intelligence will be able to solve issues in the production processes and executive robotic and production tasks.

So, how do we educate and guide the staff and students, to help them improve these complex problem-solving skills, in order to create the best working conditions for people and society?

Let's consider the well-known psychological concept of 'flow' to be a super power (as mentioned in 6.1.1.) and view it as an integral part of optimized design working conditions. One can conclude that the flow, in which humans create biological, neuroscientific linkages with ease, as substantiated by neuroscientists (see 6.1.1), is an essential part of how impulses stimulate and affect our imprinted behavioural patterns (see Csikzentmihalyi, M., 1990). These patterns are always conditioned by and related to the individual's DNA, as well as their cultural identity and experiences. Therefore, when an expert – in any particular discipline – enters into a state of flow, they can tap into exceptional powers. Their extensive experiences can be recalled in moments in which they can relax and feel confident of their skills, the same way designers can enter into **a relaxed state of creative flow. It is presumable that such moments produce the best results**. However, the ability to enter into a state of flow depends on the person's emotional intelligence and is valued and accepted to different degrees in each respective culture.

Therefore, parallel to the development in Industry 4.0 in which artificial intelligence is employed, we must also take flow – which is based on emotional intelligence – into consideration, even if it seems to stand in contradiction to Industry 4.0's cognitive, mind-led design concepts. Consequently, the flow in design will provide profitable (economic) and social benefits for the post-digital era.

6.1.3 The impact of artificial intelligence (AI) and emotional quotients (EQ) on creativity in extremely complex situations

Today, you cannot integrate AI into industrial systems without respecting and integrating embedded, *sustainable* and, in a way, moralistic effects into the innovative process. It is not only about technological possibilities on that pathway.

As the philosopher Corine Pelluchon points out, the central question currently revolves around finding a way to transform theory into practice: the act of changing consciousness into action (see Pelluchon, C., 2019). This means identifying each individual's source of 'motivation' to act that will lead to a change in human behaviour. That is what societies around the world need now – more than ever. We have to focus on the motivating factors that drive each individual's thoughts, emotions and actions. In addition, the 'affectivity' (Ricœur, P., 2016 (1950) p. 11) of these factors and moral attitudes could be considered the catalyst for changing future behaviour on social and political levels.

Today, sustainability is no longer limited to its three conventional pillars – the economy, society and the environment – as it had been since the 1970s of the 20s century. Over the last decades, the concept of sustainability has evolved into a more holistic approach, which brings the ethical actions of human beings into play (human centered design): how you treat yourself and others, the environment and life on earth (including animals and flowers), in relation to political (or social) effects.

To summarize what 'sustainable design engineering' means in 2022, we must conclude that we need both valuable and ethical action to be taken – both in digital and analogue forms. We must look at the technical possibilities as they relate to representative ethical actions. These can build upon the advantages of a globalised world and be connected via intelligent technology. And, we must demand everybody's participation.

Hence, in the year 2021, we are looking at a world which is undergoing a modification process, resulting from the digital revolution. This will lead to changes in the way we live, communicate, work and create networks between work fields – both in a practical and a digital sense. The way in which we express love, our behaviour, traditions and rituals are all changing through the use of digital technical devices. This in turn is changing our identity, influencing our way of feeling, thinking and acting (see Günthner, R., 2019; ibid: Diefenbach, S. et al, 2016).

'AI will change our creative behaviour AND the demand for ethical action'

(M.-E. Wachs, 2021, during a workshop for sketching within the AR)

Technical possibilities, the 'flow' and the complexity of digitalisation is sketching an extreme framework for 'happiness', requires positive thinking to enable creativity in exceptional circumstances.

AI is present in all of our daily living conditions: Avatars and robots are seducing us with artificial shop experiences within virtual realities, where we choose the right outfit and a second skin. 'Magenta love dessous' (see Telecom AG, 2019), created by Telecom AG, was designed to give new impulses, encouraging couples to put their smartphone aside and connect sexually in real-time. This clearly blurs the lines between reality, play, and extraordinary digitally enhanced situations.

What is the reasoning behind creating such a playful experience in a digital world that helps two individuals initiate the intimate act of love? On the one hand, it makes sense for marketing departments to create new apps and new products for the digitalised world, to serve a new consumer group: the generation of digital natives. On the other hand, it shows how a special situation – in this case a particularly emotional one – can be altered by technology and new technical possibilities.

For mankind itself the desire to be able to create freely and in democratic conditions – without fear – can be supported best by optimising education and technology. Yet, we have to ask ourselves how the working conditions in this new and modified life in the post-digital era should be, and how artificial intelligence versus emotional intelligence will play a role – especially with respect to human interaction in emotional situations. AI and EQ have a strong interconnection, not only in the normal process of designing and sketching an object, but also when creating – and deciding about – new 'complex problem solving' design concepts, with the help of the design methods 'thinking design' (Rittel, H.W., 2013)[71], or 'integrative design' (see Michel, R., 2018)[72].

Thus, it goes without saying that when using our classic design tools, we have to consider what influence AI continues to have on us as individuals, as well as on society: *'As the ethical question raised above suggest, the more digital and high-tech the world becomes, the greater the need to still feel the human touch, nurtured by close relationships and social connections. [...] as the fourth industrial revolution deepens our individual and collective relationships with technology, it may negatively affect our social skills and ability to empathize.'*[73] Like Klaus Schwab, and others, for example Martin Grunwald (see Homo Hapticus, 2017)[74] have stated over the last years, we can **recognise a dramatic reduction of cognitive capabilities, due to the use of digital technology in unusual spaces of time.** The study on 'Reclaiming conversation – The Power of Talk in a Digital Age' by Sherry Turkle (2015)[75], professor of the Social Studies of Sciences and Technology of MIT, supports this idea. **Our reduced ability to concentrate and empathise will have enormous** consequences as we enter into the next decade in the 21st century. Nicholas Carr is for economic magazines and books and visiting Professor of Sociology at Williams College in Massachusetts. Carr underlines this premise in his book, 'The Shallows – How the internet is changing the way we think, read and remember'. He states that '[...] *the more time we spend immersed in digital waters, the shallower our cognitive capabilities become* [...]'.[76]

---

[71] See Rittel, Horst, W. J., new edition by Reuter, W.D. and Jonas, W., 2013, Thinking Design – Transdisziplinäre Konzepte für Planer und Entwerfer, Birkhäuser; ibid: Jonas, Wolfgang et al (ed.), 2016, Transformation Design – Perspectives on a New Design Attitude, Birkhäuser.

[72] See Michel, Ralf (ed.), 2018, Integrative Design – Essays and Projects on Design Research, Birkhäuser.

[73] Schwab, Klaus, 2016, The Fourth Industrial Revolution, Penguine, p. 101.

[74] See Grunwald, Martin, 2017, Homo Hapticus – Warum wir ohne Tastsinn nicht leben können, Droemer.

[75] Turkle, Sherry, 2015, Reclaiming conversation – The Power of Talk in a Digital Age, Penguine.

[76] Carr, Nicholas, 2011, The Shallows – How the internet is changing the way we think, read and remember, Atlantic Books.

All these facts are playing with associations to the human feelings and expressions we saw during the first industrial revolution, when, for example, artists like Claude Monet expressed the need to reconnect with nature in order to manage the fear of machines; or depictions of changing times and living conditions, like in Monet's picture 'le petit déjeuner en plein air' 1865, where they are having breakfast in nature. During that time, 'creatives' expressed the human emotions of the day: the desire to feel again, to breathe again in nature, and to draw a contrast between naturalistic experiences and those relating to industrial, technical, and artificial experiences. Pictures played with colour, light and emotional motifs, and the field of music – another applied art – they created pictures with tones.

At that time, fine art paintings idealised or confronted reality with desires, and human emotions, in the same way that pictures in your mind create *your* world. During the first industrial revolution, the trust placed in technology contributed to economic profit and prosperity in Europe. Simultaneously, in different groups within society, machines were viewed as monsters and caused great fear of technology. Later on, Herbert Simon, an important social scientist in the design world, warned of a 'poverty of awareness', caused by the flood of information during the 1970's. In the year 1979, Simon, published his book 'Models of Thought' and described a process of thinking for creative fields and for problem solving skills. This is comparable to the current fourth (digital) industrial revolution that is in progress today and, in the future, this will again be relevant for design discussions and the ability to reflect on design and ethics and their normative character on cultural behaviour, objects and concepts. If we come back to the playfulness of 'love magenta' from Telekom AG (see above), which seduces people with the help of digital devices, tones and pictures, we must ask ourselves what it means to be aware of and 'read' these stimuli impulses 'correctly'.

When we look at all of the technology related to the digital revolution, the 'patterns of pictures' in 'deep fakes' will play a significant role in the next cultural revolution. Designing 'deep fakes' with the help of AI, and with an **understanding of the influence that coloured dots, light and shadow can have when using software patterns, which helps create the illusion of authentic pictures, whose content is seductive and dazzling at the same time**. Yet, we must be aware of illusionary images, such as 'deep fakes' of video footage, which have had a significant influence in the US, like is described in the following.

In order to help us understand how the experts work, the article 'Fälscher auf Speed' (Welt am Sonntag Nr. 49, 8. Dec. 2019) researched the process of how patterns and pictures are constructed to imitate reality, with the help of natural and artificial light in the image. Jacob Hesse mentionend in 2019, that natural light always creates patterns, for instance, which do not exist within image constructions created by software via AI.[77] Hesse, an IT forensic expert, is seeking to differentiate between truthful content and artificial design realities. We must accept that pictures and videos are influencing and manipulating the way human beings think and act. Furthermore, experts are expecting an optimisation of the process, but at the moment we can still assume the following: **the image that is**

---

[77] See Hesse, J., in: Welt am Sonntag Nr. 49, 8. Dec. 2019, translated by M.-E. Wachs.

designed by AI is only as good as the expert's skills, who has created the pattern of the image. Does that mean that the more we know, the better the artificial product will work? Not necessarily. It is a question of cultural education, research interest and the next generation's skills in using 'hands on design tools' and digital devices. Yet, in the future, the skill of creating trustworthy content, that creates 'realities' – and ethical value will also be highly relevant. In conclusion, it can be said that 'extreme problem-solving skills' (see Simon 1979, ibid: Schwab 2016) are one of the most important skills of human man kind in the future.

6.1.4 Exceptional circumstances and Systems Oriented Design (SOD) in a post-digital century

So, the question now is, **how** would we like to design the next 'realities' in the digital era? How can we use AI to create new processes for the industry, for life and for cultural behaviours (rituals, traditions, loving interactions), when a nation is falling back on the 'cultural memory' (Assmann, J. 1997 – German: Kulturelles Gedächtnis), reflected in trained behaviour that has developed over the past centuries? Generations of human beings are connected by means of these cultural minds, which create their understanding of culture – their cultural identity. The development and imprint of a cultural identity is constructed through a continual process that trains people's habits (see Schiffer, M. B.; ibid: Hahn, H. P.; ibid: Assmann, A.; ibid: Assmann, J.). By mediating these habits over time, the cultural system is an ever-evolving element of society that is passed from generation to generation. In these circumstances, **time and emotions have an important effect on whether people take action, and this is important for creating a cultural identity** and cultural heritage.

When the Financial Times Weekend, from 14th-15th December 2019, named the year 2020 the 'Age of the Avatar', it reflected society's current focus on a world in which narration and fiction are mixed, by combining real technical representation with the help of virtual celebrities: '[…] *digital celebrities are only the most visible aspect of a growing trend. As the technologies behind them become cheaper and more accessible, you might find yourself listening to music released by a digital avatar, interacting with one in a convenience store or dealing with them in your workplace. Soon, you may have one of your own.*' (Bam, B. 2019, FTWeekend, p. 6 Life & Arts).

Virtual avatars and YouTubers are creating a new kind of reality, and way for humans to connect with one another, and further examples can be found in art objects like 'Bina48'.[78] We are pursuing 'synthesising software', as well as synthesised objects and ideals, all of which are influencing our feelings and actions. Therefore, it is necessary to gain more awareness of authenticity now. In addition, we must recognise the ethical impact this is having on complex problem solving in the digitalized world. Are 'Instagram avatars' creating trends, or do we as designers create these trends as well as the avatars?

These trends are creating images within an experimental field, playing with demands and the market's technological possibilities. It is obvious that the designers as well as the consumers, are both highly responsible for these design trends, which are creating a strange and fiction-based world, using AI. Artificial models in businesses were the trend in marketing tools from 2017 to 2019. These communicated in a transparent but experimental way: for example, to make statements about 'aesthetic habits', the attitude towards 'queerness' model marketing, and other business models (see 'Shudu', Balmain, 2019 created as prototype for a black woman; ibid: 'Miquela', created as a model, representative for queerness in 2018). So, AI applications are changing how we perceive reality, by creating new ideals with respect to the beauty ideal of this unprecedented period in time, and by exploring new business models that incorporate immateriality, while presenting facts as well as fake news that is being designed at the highest possible speed.

In this century of new digital patterns, Meadows's Systems Oriented Design (SOD)[79] is helpful for managing these complex problems and developing everyday solutions and business concepts. At the same time, it could potentially be implemented as a design method in creative fields in design, alongside the 'flow'. However, in contrast to the concept of flow, SOD is a tool that allows you to think in terms of systems (see Gulden, T. and Støeren Wigum, K., 2019)[80], and it is used for analysing and exploring research and design – as discussed in the next chapter. The flow is an important tool for creating in an active way, but not for analysing, or exploring thoughts – when you are in the flow you just do, without thinking. This can be an act in 'exceptional circumstance'.

[78] See Antonelli, Paola / Tannir, Ala, 2019, Broken Nature – XXII Triennale di Milano, p. 181 ff, art project from 2014.

[79] See Meadows, Donnella H. / Wright D., 2015 (2009), Thinking in systems: a primer, White River Junction: Chelsea Green Publishing.

[80] See Gulden, Tore and Støeren Wigum, Kristin, 2019, Keeping up with complexity, in EPDE 2019, Towards a New Innovation Landscape, p. 407 f.

6.1.5 A naïve view enables creative flow in extreme working conditions – using artistic skills and equally valued humanities alongside STEM in advanced design education

All the previous examples of digitally created artificiality, show the immense, wide-spread influence of artificial intelligence. **They all deal with the complexity of our habits and the design process, which always relates back to the way humans feel and think.** From an economic perspective, the aim is to launch optimized products on the market that offer a high level of technical performance and bring the greatest amount of profit. From a psychological and cultural anthropological perspective, we have to think about and question the relationships between 'people and things' (Schiffer, M. B., 2009; ibid: Küchler, S.; ibid: Assmann, J., et al), as well as 'people and concepts' (Wachs, M.-E, 2008, ibid: Wachs, M.-E., 2018). This will help us incorporate the wider perspective of design theoretical evaluation, which brings us back to the hypothesis and poses the questions: **How would we like to create new habits with the help of AI – and emotional intelligence – in terms of 'ethically engaged' behaviour. If we ask how to create these new habits, we must automatically also ask: what kind of a design methods should we use in the creation process? And, what kind of habits do we want to promote (– to design)?**

While we must consider what habits to shape, it is also important to look at what sparks us to act and to move forward. **The term emotion**, a translation of the Latin 'e movere', means 'moving out of something and into another', and implies that emotion leads to a specific action, which comprises a 'drift to act'. The clinical psychologist Goleman explained this in his book called 'Emotional Intelligence – EQ'[81] (Goleman, D., 1995).

As we have seen in the examples outlined in chapter 6.1.3 and 6.1.4, and as psychologists have established, pictures are visualisations of human thoughts, feelings and actions: here, the cultural imprint (not only that of childhood) as well as the impulses/triggers and conditions surrounding the human are definitive. Fear is one possible trigger for action, but a comfortable atmosphere, a positive surrounding, or enabling a person to feel happy and free may serve as an even 'better' trigger point for creativity and problem solving. All of these considerations are elements of each individual's 'emotional intelligence', and promote connectivity to people and to systems. (see Hüther, G., 2009, p. 81 f, ibid: Schwab, K., 2016, p. 101 ff, ibid: Goleman, D., 1997). The neuroscientist Gerald Hüther mentioned how trigger points are impulses, signals or disruptions, within a familiar pattern to act. During the 1990s, Mihaly Csikzentmihalyi referred to the 'flow' as an

[81] See Goleman, Daniel, 1995, EQ – Emotionale Intelligenz, dtv.

Note
Hüther, Gerald, 2009, Die Macht der inneren Bilder – Wie Visionen das Gehirn, den Menschen und die Welt verändern, V&R.

optimized creative space, which functions as a trigger point for impulses, whereby the flow is allowed to take place in a moment of boredom or fear, (Csikzentmihalyi, M., 1990; ibid: Goleman, D., 1997, p. 121.)

Although fear can serve a trigger for flow, freedom of creativity in extreme situations can be optimized by creating a stress-free environment: for the best performance, we need a design method that provides us with a stress-free space, enough time, and the security of our expert skills, which we have improved over the course of many years. A design method that is developed to create new habits would integrate the concept of flow, as this is a good way to generate ethically engaged and (trustworthy) cultural behaviour, and, ultimately, to evaluate new concepts supported by AI. This approach raises many questions that can be answered with the help of other disciplines, such as behaviourism beside technosciences, as psychology and philosophy, regarding ethical values, by design theory and by sociology and in 'cooperation' with engineering, mathematics and physics. In Germany, or in countries with a Western cultural background, answers can be found more easily in the so called 'soft disciplines', rather than the 'hard' ones that quantify the measurable 'direct' economic factors of the so called 'STEM' subjects (STEM = science, technology, engineering and mathematics, in Germany 'MINT').

However, these soft disciplines, like philosophy, ethics, sociology and design theory, are not highly appreciated in Europe at this time, which prevents them from having optimal financial value. Yet, as the answers that these disciplines can provide are relevant to all societies and nations, it is in fact a political issue that is important for the cultural education of diverse social groups – and furthermore to geopolitical shifts.

In this day and age, it is obvious that designers and researchers are not the only ones responsible for creating democratic and sustainable cultural systems. Therefore, it is very encouraging to see how a growing number of foundations and entrepreneurs are also getting involved and promoting more soft disciplines and educational programmes.

The 'Humanities and Social Change – International Foundation', founded by the entrepreneur Erck Rickmers, in 2018, and continued in 'The New Institute' (-Foundation gGmbH), is one example of how to take responsibility. Still, we need more heroes to get involved in cultural education in Germany and Europe, because this is our most valuable capital: knowledge and humans.

6.1.6 Flow, ICI (integrative cultural intelligence) and happiness in extreme and creative situations – moving towards more humanities in design engineering

This research chapter encourages the need for cultural education in soft disciplines and extremely stress-free, secure spaces for getting into a design flow, allowing for the fluid development of human interaction with objects and concepts and solving problems in complex systems. **Artificial intelligence is forcing the next step in industry and cultural behaviour, whereas emotional intelligence could design a new artistic intelligence.** The latter requires the support of the soft disciplines – defined in Germany as ranging from fine arts to design theory, to ethics, sociology, and philosophy. **This kind of new artistic intelligence**, not only integrates multidisciplinary anthropological skills. It **can also train cognition ability, concentration and complex problem-solving skills, because it generates a deeper understanding of abstract phenomena**. It is evident that, besides technically optimized systems and AI, humans and sustainable life on earth both rely on emotional **and** artistic intelligence, to develop an 'integrative cultural intelligence' (ICI, Wachs, M.-E. 2020).

Within the scope of cultural education, ICI enables all people to understand abstract signs and complex systems. This is achieved by using inner impulses and patterns, which incite taking action (see Hüther, G., 2019). This new Integrative Cultural Intelligence provides orientation and helps train new behavioural patterns and rituals that lead to new sources of motivation. The resulting new rituals, combined with the motivation to act – again referencing the philosopher Corine Pelluchon – are what will help form the cultural mind(set) of the future. The motivation to act responsibly, is the product of each individual's cultural imprint on how they think and feel, which is influenced by the cultural memory of the society they live in. **Ergo, cultural education is the source for living a respectful and ethical life on earth – with AI.**

Needless to say, people need a new Integrative Cultural Intelligence to solve design problems in a human-oriented way, and the most optimized way to achieve this is through 'design in flow'. As we have seen, the flow – a state in which creating and designing is fostered in stress-free and optimised 'spaces' – functions as a super powerful and should no longer be wishful thinking.

In addition to **ICI, there is another aspect to enabling people to take action**. In the book 'Cultural Intelligence: Individual Interactions Across Cultures', by Christopher Earley and Soon Ang (Earley, C., et al 2003), they describe how the best way **to improve one's ability to 'adjust' is to lead a cross-cultural team**. However, the influence of AI on changing working conditions means that a new cultural intelligence will provide the people with 'change-enablement' – not only 'change-management'.[82] **Change-enablement refers to the individual's own motivation to act.**

This can be used in conjunction with a modified ICI, thus providing a higher quality wealth evaluation system for sustainable culture than (e)valuating with quantified models can. The combination of using this and **'art' as a trigger could be beneficial for the economic system, by producing even higher economic profits with the help of 'indirect-profitability'**. In other words, implementing real cultural education – triggered by art, artistic disciplines and soft sciences – profitability would be several times higher than the direct profitability, that is derived from the hard disciplines, such as engineering and mathematics – the STEM fields of study.

A truly resilient and sustainable economic and education system has to be the prerequisite for designing engineering landscapes on the one hand, and for economic prosperity on the other. Yet, what about the extraordinary circumstances related to the 'creative economy'? What are the specific qualities needed when using artificial and cultural intelligence in extremely limiting working conditions, with factors related to time, energy and resources that pose strong impositions?

Would it be best to cultivate design flow, by providing experts with enough time to work in stress-free spaces? And will the use of soft disciplines form the foundation of holistic, interdisciplinary and cross-cultural education? If so, then let us return to the **idea of profiting from a 'naïve view of the otherness'**[83] (Ricœur, P., 2016, p. 191, translation: Wachs, M.-E.) – or in other words, the idea of culture in embedded System Oriented Design and 'living with complexity' (Donald Norman).
If using artificial intelligence has a positive effect on seduction – by boosting genuine appreciation rather than doing damage to somebody or society – then let's be seduced by the flow, into designing for good.

Knowing that we continue to work with artificial intelligence, we **have to make sure we nourish our cultural intelligence, in order to develop innovative and sustainable design solutions** in extreme circumstances – like the ones we are finding as this digital world begins to evolve.

Human connectivity and integrative design is just as important as being aware of how future cultural behaviour will be created, with the help of artistic and artificial intelligence and integrative cultural intelligence (see graphic 07, AI – EQ – ICI by Wachs). Creating our future with AI, EQ and ICI means that researchers, users, designers and other experts, like politicians, must take responsibility. This will have a positive impact on both artistic and artificial intelligence, created by and for the designers, creators, children and entrepreneurs, and also integrates all of society's stakeholders. So, let it flow and make sure to create design engineering education landscapes that include stress-free spaces, for reflective and relaxed working conditions.

---

[82]
See Günthner, R. et al, 2019, Hirn 1.0 trifft Technologie 4.0. Springer, p. 88, ibid: Sander, Constantin, 2010, CHANGE – Bewegung im Kopf, Business Village; ibid: Dueck, Gunter, 2010 et al.

[83]
See Ricœur, P., 2016, Das Willentliche und das Unwillentliche, Wilhelm Fink, p. 191, translation Wachs, M.-E., 2020.

Graphic 07
Trigger points of advanced management of knowledge: artificial – emotional – integrated cultural intelligence in design, Marina-Elena Wachs, 2020.

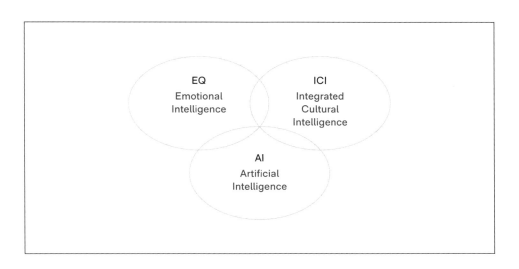

6.2 Virtual Interactive Design (VID) – shaping digital and ethically correct spaces for creativity and its benefits to stakeholders

Keywords
AI and ethics in interdisciplinary, digital teaching and learning formats; case studies in design engineering; design theory and the history of design; significance of technological history; ethics in design; interlinked collaborative learning lessons – 'Cross-Cultural in Europe', virtual interactive designing formats and ethical questions.

This chapter describes a project, first conceived in 2018, for digital and ethically correct 'spaces for creativity', within interactive university learning landscapes and economies. The project is continuing to evaluate the experiences gained over the past years of digitisation, thereby learning how to create future learning landscapes in design and engineering – and beyond. What do we need to shape ethically 'correct' compliance, within Europe and around the globe, that will help build 'digital competencies' for using digital tools – not only with focus on design engineering and creative industries.

6.2.1 Digital and ethically correct spaces for designing – what role does interactivity play?

The long-term project 'Digital ethical correct design spaces in the interactive university landscape and economy' offers an innovative learning opportunity – primarily for master students, but also others taking the **module: 'Virtual Interactive Design (VID) – Cross-Cultural in Europe'**.

The concept sketch was formulated from a design engineering perspective. Here, design drafting methods, as well as design theory models and methods, shape the basis for didactic tools used for interdisciplinary, integrated learning. In addition, 'interactive designing' has evolved since the 1990s and now means fewer digitally networked tools, as compared to metaphorically interactive design processes.

The innovative learning offered by 'Virtual Interactive Design' (VID), combines self-learning via online courses – called 'MOOCs' (massive open online courses) – as well as blended-learning practical courses that focus on events with the following objectives:

1. Students train how to use artificial intelligence in a reflected, ethical manner – machine-based learning.
2. Two newly designed MOOCs impart basic knowledge of AI competence, in special subjects (e. g. history of design/technology), in order to fulfil the third objective.
3. Here step 2. furthermore enables interdisciplinary, culturally networked interactive learning and design by means of project-related MOOCs.

In the VID project, students learn to collectively train their AI competencies at peer-level, as well as with external actors (collaborating enterprises, other experts), while using different digital teaching/learning settings. In addition, they use methods that help them reflect upon ethical usages. At the same time, these future experts have the opportunity to train their AI competencies. Ideally, these projects involve 10 to 20 participants. Theoretically it is possible to scale it up to 100 people – though it is questionable whether that is a sensible group size, especially when building 'complex problem solving' skills within the European region and through interactions with stakeholders from businesses, politics and research. This can be achieved through OERs (Open Educational Resources), which enable individuals to practice self-learning, self-reflection and evaluations in peer assessment procedures, which translates into lifelong learning for everyone in society. The Covid-19 pandemic has taught us, in a very short space of time, what is possible from a technical point of view. **However, at the same time we have to perform a precise didactic evaluation, to determine which consequences this will have for new learning – and interactive design – formats: combining the best 'from blackboard to tablet' in interlinked learning – in ethically correct spaces.**

6.2.2 The state of research in digital learning landscapes – What are the core objectives, questions and operational requirements for implementing VID?

The aim and core objectives are *to create digital and ethically correct design spaces in the interactive university (learning) landscape and economy.* In terms of teaching in the industry and at universities, the change process in the field of 'design engineering' began no later than 2016 (see conferences in the UK, including EPDE; Nordic countries, including ICDC; WEF Davos). This has been recently been accelerated by the contemporary fields of artificial intelligence (AI) and sustainability, which have brought about massive motion: As outlined by the World Economic Forum (WEF) and Klaus Schwab's publication 'The Fourth Industrial Revolution' (2016), and other publications and events around this year, a new awareness of the use of artificial intelligence has increased dramatically.

For us designers and university teachers, this means creating a new learning environment characterised by students learning together with industry experts and research institutions, including (international) alumni in networked digital learning formats. Above all, legal and ethical questions have to be clarified: How can artificial intelligence be used in the design process and how it should be used in a socially responsible manner – in addition including culturally diverse participants? However, there are also major reservations on the part of certain disciplines, and on all three levels of education (school, university, industry). This applies to design in general among the younger generation of students, who are questioning how and whether they can use methods of artificial intelligence to create designs within their studies, as well as in the industry. **It is important to break down this bias through integrative methods and new didactic learning formats.**

The term 'design spaces' does not only refer to physical spaces that are equipped with digital networking tools – appropriate hardware and software – that are to be as widely used as possible, and optimally in divers combined formats with Open Education Resources (OERs). The question is also, when does it make sense to employ alternate micro-learning content with the use of MOOCs and blended learning formats, and how can the younger generation be motivated to make this change?

In a figurative sense, the term 'design spaces' also refers to the creation of a legally correct arena in which we can move and train a new generation of digital natives. We must be aware that ethical values regarding the use of artificial intelligence are also at stake here. **On the other hand, 'Gestalträume' in design are meant to be thinking spaces, and**

immaterial spaces that sustainably incorporate emotional intelligence – thereby also nurturing competencies such as empathy, participation, integration and inclusion. In addition, these spaces represent metaphorical models for abstract thought. The greatest challenge and most important requirement of these future experts will be their 'complex problem solving' abilities, within the post digital-era[84] (see von Weizsäcker, E.-U., et all, 2018) – that includes an holistic education and understanding about abstract phenomena. Moving forward, both basic technical knowledge and networked fields of work will be required, to help solve social and economic problems within interdisciplinary, cross-cultural fields of development. This could be best supported through teambuilding and coaching, systems-oriented-designing and design thinking methods, for example. The shift from linear economic fields to circular economies – and the decentralized management process that will follow – do not only mean that simple production actions will be performed by robots using artificial intelligence, and that humans will lose such jobs. It also means, it **will be essential** to be able **to read and understand abstract models, and large industrial and economic abstract systems, in order to initiate the next steps of development**. The Internet of Things (IoT) and Internet of Production (IoP) are showing us new ways to use AI-based, technologically optimized solutions. But the students' question remains: How these can be applied in an ethically correct manner to develop the solutions of the future?

In this context, not only competences in the field of 'analogue and mixed digital processes' will be necessary. In particular, a rethinking of Donald Norman's model of the 'human-machine interface' (1988), in which he showed that this interplay would be evoked by the digital revolution. Questions of how to deal with 'things' versus 'processes' has become more important in terms of 'cultural behaviour,' which has anthropological and system-oriented significance. For design engineering research done in the Nordic countries and the UK – which also took empirical studies from China, Australia and the USA into account – a design-theoretical approach, such as systems-oriented design, has been used in recent years. This was the preferred approach for interdisciplinary project-based learning formats by the author Wachs, to promote 'complex problem solving' competence, among other things.

It is striking, therefore, that the parameters of digital learning landscapes and the use of AI technologies are still managed in an unsophisticated manner: Design learning and development methods continue to be widely based on design-theory, as can be seen in the design methods of design thinking, integrative design, as well as UX design or social design. In contrast, *'interactive design is based less on digitally networked tools than metaphorically interactive design thinking'* (Wachs, M-E., 2019, EPDE Conference Glasgow)[85].

The biggest change in dealing with the 'physical' object-like world of things, in the Internet of Things, is that we must face the emotional world: Many ethical questions arise when people want or need to keep up with change. Artificial Intelligence (AI) faces Emotional Intelligence (EQ) when we look at how people use and manage objects, processes and ideas. **Emotional intelligence and questioning ethical value in relation to the**

---

[84] See von Weizsäcker, Ernst Ulrich / Wijkman, Anders (ed.), 2018, Come On! – Capitalism, Short-termism. Population and the Destruction of the Planet – A Report of the Club of Rome, Springer.

[85] See Wachs, M.-E. / Hall, A., 2019, European Driving Range – Innovative Landscapes For a Tangible, Non-Hierarchical Learning Space Within a Material and Immaterial Togetherness, EPDE 2019, Towards a New Innovation Landscape, p. 498 ff.

use of open-source formats have not yet been sufficiently considered. Equally, ethical considerations, relating to the sustainable use of AI, are currently being heavily discussed from a philosophical standpoint (e. g., Pelluchon, C., 2018, ibid: Precht, D., 2020). However, this must also be viewed from the design engineering point of view of – which will subsequently affect the habits and conduct of society.

The ethical – and thus sustainable – use of Artificial Intelligence and Emotional Intelligence in the human-machine model impacts the economy, industry and trade, as well as the creative industries. It also plays back into the educational frameworks of the future, and thus into the university landscape on an international level. Above all it represents 'value' for the individual and society.

As design researchers, it is our responsibility not to leave the 'digital natives' alone to manage AI's vast technological possibilities. Together with the students and other actors outside the university – such as, representatives from professional and life science fields, as well as industrial and social experts – we must reflectively approach the active process of shaping these 'spaces': This begins with the provision of analogue and digitally-based learning settings, the collaborative design of MOOCs and learning nuggets, and by training the ethically correct management of 'design spaces', which should not be limited to simply asking critical questions about the use of AI-based technology and data rights. It **is not so much a competition between 'man and machine', but rather an ideal and the complementary interplay and meaningful use of the players**. The question that should be discussed in a holistic framework is: Can digital competence be separated from AI competence?

Key Objectives

In order to highlight the core objectives in formulating the framework for future design engineering, we have to look at the macro and micro perspectives from multiple viewpoints: i.e., considering both the digitally ethical design spaces in the interactive university landscape and economy, as well as learning through Virtual Interactive Design (VID).

Therefore, if we initially restrict things to the cultural space of 'Europe', which serves to respect the European Rights Agreements at the core of ethical behaviour in terms of digitalisation and sustainability, we have to consider implementing the following:

- new interdisciplinary digital learning platforms for students and teachers, together with industry and business experts, covering all disciplines
- analogue learning settings in combination with digital teaching-learning settings: learning landscapes together in Europe, via digital tools but with a human-centered basis
- innovative future formats for transparent communication – 'Open Education Tools and Research' used in a sustainably integrative manner and developed together with different stakeholders – including students; a basis to provide them with a starting point (MOOC basic course), upon which interdisciplinary interaction – both analogue and digital – can then be built, thereby promoting 'self-confidence' in their own competencies (not only among students)
- training for AI-based competence means Artificial Intelligence (AI) AND Emotional Intelligence (EQ): Robots' physical intelligence acts on the basis of artificial intelligence, yet still needs the emotional intelligence of human beings in order to integrate the 'products' that can be understood as both object-like items as well as services used in everyday life
- more joyful learning landscapes in the future, by reducing prejudices and not reinforcing fear; a playful (and experimental) approach creates an opportunity to become a problem solver, by means of experimental design methods, design theory and VID
- at the micro level, this means, first and foremost, designing and testing a blended learning format: using integrated learning formats with a media mix means collaborating with the consortium of the submitting parties and optimising them with MOOCs; the consortium should comprise professors and media didacticians, IT and AI specialists (both internal university staff and external partners), philosophers and psychologists, as well as other entrepreneurs and representatives from business and society (numerous legal and ethical questions will arise, for which external experts can also be consulted) – this is the advantage of multi-perspective and multi-field solutions.

Research Issues

It follows that the (mostly didactic) research questions are empirically and hermeneutically substantiated by Marina-Elena Wachs's interdisciplinary (cross-cultural) project:

— How do we unite the values of analogue and digital design skills in relation to our own subject, design / engineering? We must consider that the training of AI competence must not only take ethics into consideration, but also provide a sustainable way forward for future educational programmes.

— How can innovative, networked learning spaces – and thus design spaces – provide added physical and immaterial value for education and the training of experts in the post digital era (i.e., for industry, economy and society at large, and for design/engineering in particular)?

— How can AI become a learning model for 'virtual interactive design', creating ethical digital design spaces, which can be sustainably transferred to various disciplines (subjects)?

— How to take benefit by integrating synaesthetical trainings into interactive design methods?

— How could we motivate the next generation, to formulate more critical questions to the system?

Operational requirements for implementing the innovative design learning model – like a modular 'virual interactive design' (VID) – offer are as follows:

1. software compatibility, blended learning environment – media technology, networking

2. establishing a virtual organisational structure that is respectful of data rights – needed for creating user trust, while keeping the financial budget in mind

3. creating MOOCs (massive open online courses) with different learning and testing elements for self or group learning and evaluation, combined with implementing basic courses, in addition to an interdisciplinary project

4. switching back and forth between analogous learning settings and interdisciplinary digital classrooms

5. testing application scenarios: students of different study programmes and levels should do this by addressing an overarching social topic; they should implement interdisciplinary, cross-cultural testing from many perspectives, and explore AI's possibilities, while training to ensure its ethically correct usage; such training includes an open-minded exchange and 'critical reflection' of AI in the learning application format

6. later, also developing practical digital supplementary modules – elective modules (in combination with: VR, CAD, 3-D, MR, AR and future tools to come, e.g. a 'textile tab' (see Wachs, M.-E., 2021))

7. evaluating transfer possibilities of design engineering models to the following disciplines: mechanical engineering, all design branches of applied arts – from fashion design, textile design, interior design, architecture, through to culture and anthropology

8. evaluating constantly demands for modification the interactive designing / learning system by design didactic findings.

6.2.3 Perpetual cross-media learning tools in design engineering for talents, experts and civil societies – potential transfer of benefits to other disciplines and its cultural significance

The application of innovative, digital and ethical design spaces will benefit students of all disciplines – from Bachelor and Master to PhD programmes – who form the target group. During the development phase, and the creation of the AI-based learning formats and interactive learning tools, teachers and experts from the hardware development and software industry, consulting companies, industry representatives and entrepreneurs, as well as foundations and council members will all form the interactive layers with which they and the students work within together. In the end, the model of 'Virtual Interactive Design Cross-Cultural Europe' can be transferred throughout Europe's university landscape, and also to learning models for civil societies, or experts from business and industry. A 'perpetual' learning model means one which is developed and evaluated again and again – thus a collaborative modification of the tool's elements every two to five years! The model survival will be ensured by caring about the significance of the valuable, morally correct usage of designing tools for both parties: education and industry.

In terms of content, this means that with this project, we are developing AI learning solutions, in accordance with the various stakeholders, for the professional fields of design, design engineering, for all design sectors and for the applied arts. Furthermore, it will later develop into a multi-perspective and broadly transferable approach that can be implemented in many fields – from mechanical engineering to anthropology. In addition, it aims to promote the critical and reflected use of AI and AI-based technology, for example, in the selective implementation of image recognition software and image search masks, while respecting image rights. By means of digitally networked learning landscapes and the use of MOOCs, together with innovative collaborative learning systems, a level of lifelong learning is simultaneously made possible, thus ensuring everyone's participation in cultural education.

With the help of the described and further on developed model (VID, 2018), we could develop solutions not only in the design engineering field, but also for building interdisciplinary competencies and addressing the need for a value-based, sensible and general use of AI. This in turn also affects the following areas in particular: ethics, design/engineering and ethics, design/engineering and cultural studies, including every day applied design solutions. Even politics and sociology are affected by this shift.

To answer the question posed at the beginning of chapter 6.2, digital and ethically correct *'design spaces in the interactive university landscape and economy'* must be developed for cross-disciplinary actions with societal value. The economic benefit of the VID model provides resilient long-term economic and cultural value in sustainable design engineering models – across multiple disciplines. To reach the next level in social responsibility, it should be the aim to further develop and incorporate the United Nations' SDGs (Sustainable Development Goals).

6.3 In *consideration* of a holistic design egineering education – integrated cultural intelligence

6.3.1 Encouraging the Ministers of Education and Economy to value the social impact of fine arts as much as STEM

Keywords
Integrated Cultural Intelligence (ICI), consideration of holistic design education; materialising the immateriality; researchers' responsibility; more humanities in design; design doing in pre-school.

The benefit of a valuable and 'considerate' design education, lays in transferring holistic knowledge into designing concepts and in 'materialising thoughts'. The word **'consideration'**, as the Latin term suggests, is about the philosophical significance of decision making as well as thoughts that show respect for the value of other things in a wider sense (see Pelluchon, C., 2019, p. 36 f.). Keeping a bigger picture in mind, it can be interpreted to include the expression of **well-considered** design concepts with the help of a visualising model, which requires a basic understanding of abstract signs and media. Integrative design and virtual, collaborative design models could be very successful, if members of a team who have an equal appreciation for education and contemplate and embrace the benefit of each other's approach as they work together.

This research study is also encouraging the Ministers of Education and Economy to value the social impact that fine arts, design theory and other *soft disciplines* have on smart solutions, alongside STEM. In the process of defining the characteristics of a culture-based, valuable and holistic design education, a wide range of (design) languages, as well as societal and individual habits are shaped. Ergo, it would be 'considerate' to implement holistic design education programmes, from pre-school all the way up to non-hierarchical learning systems in extended vocational training. After all, this is how the future in multimedia and multi-perspective design engineering spaces should be 'sketched out'.

Therefore, let us accept that valuable design education is well worth the 'consideration', as is the transfer of holistic knowledge into design concepts and the 'materialisation of thoughts'. 'Consideration', as it relates to the philosophical significance of decision making – in that form like Corine Pelluchon describes as 'considerare' –, also expresses the idea of respecting and thinking about and taking care of others. It focuses on the philosopher Pelluchon's description in 2019, within the frame of 'humility' and in relation to the individual's motivation to act. Humility requires us to distance ourselves from our own needs, particularly when it comes to serving sustainability. This is an important motivational factor for actually taking action – as opposed to just thinking, speaking, and discussing.

In her book 'Ethik der Wertschätzung' (ethic of valuation), Pelluchon argues that we need to speak about what value means, and that the relationship to others and the world around us is proportionally associated to our self-worth (see Pelluchon, 2019, p. 36 f).

This futuristic 'playground' in design engineering education will be encouraged by philosophical augmentation towards more 'corporeality and impulsiveness' (see Pelluchon, 2019, p. 26 ff; 128 f), and will enable the mediation of knowledge at the earliest age possible, in a playful and worthwhile manner. The **value of time during the 'artificial engineering design' process emphasizes the worth of the 'interdisciplinary and cross-cultural act'**[86] **(Wachs, 2008) that involves combining art + design + engineering + humanities**, more than ever before. A 'naïve' view of 'otherness' (Ricœur, 2016 (1950), p. 191), and the resulting tolerance, could lead us to new design methods and innovative benefits for society and economy in special, by interdisciplinary interactive projects.

The following research results incorporate research studies and education experiments from 2019, which were performed in a comparable manner at various universities in Europe. They evaluated hermeneutic studies of university publications and trade literature, ultimately reflecting on these results with the help of expert interviews in the fields of design and fine arts. A holistic design education in the post-digital era is essential for promoting the necessary respect for researchers, and fulfil the responsibility of creating a worthwhile design engineering landscape at the higher education level, in parallel to Industry 4.0. As such, the meaning of the term 'industry' will continue to change, and be modified by the ongoing process of digitalisation (see M.-E. Wachs, 2021).

---

[86] See Wachs, M.-E., 2008, Material Mind / Materialgedächtnis – Neue Materialien in Design, Kunst und Architektur, Dr. Kovač.

6.3.2 Scientific disciplines still have a higher value – more focus needs to be placed on design engineering education models and pre-school education

Integrative design (Michel, 2019), design thinking (Plattner, 2009) and systems-oriented design (Maedows, 2008) – also in further developed forms like 'keeping up with complex systems' (Gulden, Støren Wigum, 2019) – are methods that could be very successful if all developers in a team work together, as long as they appreciate the value of each other's education and work on a level playing field. They need to understand and accept the benefits of each person's approach and the perspectives of others. However, up until now, financial support for cultural education at schools in Great Britain, for example for music, sports and arts (The Herold and other newspapers, September 2019), and in Germany, is very limited. In recent years, since the Bologna Process, a higher value has been placed on the STEM subjects, and, as a consequence, they have received more financial support than the humanities in study programmes across Europe. Yet, we know that abstract thinking and connected areas of the brain, which **work together in a symbiotic and synaesthetic way, are essential for understanding language and complex systems. This means the 'hierarchical value' given to courses, programmes and disciplines at universities and schools has had a negative impact on the benefits to society** provided by design engineering solutions. This will continue unless changes are made to promote a more long-lasting artistic, creative mindset that begins as early as pre-school.

Since 2016, and as claimed the chapters before (see in detail chapter 6.1), though complex problem solving is not a new phenomenon in design. Nevertheless, the tasks people will have during and after the fourth industrial (digital) revolution are changing, and by the end of this revolution the number of problem-solving tasks will be greater than ever before. This is due to the fact that artificial intelligence will complete tasks in production processes, including executive tasks. Thus, it is obvious that quantifying numbers and measurements will be relied upon to yield direct economic profit, which will be achieved with the help of STEM study programmes. In consequence, there will be a shift in working conditions, moving towards more automated production processes performed by robots, and processes operated by artificial intelligence. As a result, *quality* measurement tools performed by humans will come into focus. Therefore, it is undeniable that the indirect economic profit provided by cultural education – triggered by soft disciplines like philosophy, design theory and fine arts – serves as the precondition for the complex problem solving needed in the future.

6.3.3 Design methods and cultural design codes in Europe's education – case studies in 'Textile Engineering Education Strategies in Industrial Design'

As mentioned, this study is encouraging the Ministers of Education to place a higher value on the social impact of the arts and design engineering and their contribution to better future life conditions: humanities, design and art must be valued just as much as mathematics and physics, by focussing on incorporating cultural education into early learning programmes and offering children playful hands-on experiences.

The case studies look at and question the differences between German Design Engineering study programmes that focus on fashion or textile, and, for example, the Scottish study of Sports Engineering programmes (at Strathclyde University)[87]. It thereby aims to provide an open-minded evaluation of cultural diversity:

First, we can look at some German educational processes in design prefer (material) performance-driven design processes for sports related garments and running shoes. There is a strong focus on yarn performance and textile structures – within the framework of 'Textile Function Management' (Wachs, M.-E., 2021) – which impact compression, breathability and air circulation, for example. The way in which this design process generates innovative design solutions **for future industries is to implement nonlinear production processes** – in other words circular and decentralised processes. With the help of body-mapping, the warp knitting baselayers are built up with the help of various material and pattern-based substrates. It is about creating smart sportswear **that serves and supports the client in specific (athletic) disciplines**. We see this in the products presented by enterprises, such as Adidas, Lulu Lemmon, Odlo, On Running or Patagonia for example, which showcase more than just high-performance textiles: they utilise super smart textile and sports engineering, to support the body in every possible way.

Sports design engineering for disciplines like running, has been defined and optimised by computer-added systems for the last decade. However, now it is important to consider garment requirements and affordability for high-performance objects that contain technical textiles, such as those needed for Paralympic sport disciplines. These aspects should be incorporated at the beginning of the design process. Furthermore, the use of technical textiles and the challenges they bring not only apply to the field of sports engineering design – but in fashion and clothes.

---

[87]
See Strathclyde University, (2019), Sports Design Engineering – Why this course?, available: https://www.strath.ac.uk/courses/undergraduate/sportsdesignengineeringbeng/ [Access 2020, 04 Jan.] https://www.strath.ac.uk/engineering/designmanufacturingengineeringmanagement/

Note
Some of the designs are based on the high quality of the textile fabrics, developed 'Made in Germany' by Penn Textile Solutions, Paderborn – Germany – thanks to our cooperation partner not only at workshops 'Textile Pop 2019'. https://www.penn-ts.com/en

Other future social problems, such as migration and integration, will compound these challenges, as affordability becomes an even greater issue. The high-quality solutions found in sports textile engineering – like the extremely powerful technical textiles and manufacturing techniques used for prosthetics – will become a part of everyday life, such as wearing body supporting 'design tools' every day. In our society, which has a strong focus on wellness and health culture, enhancing body performance technology will be integrated into wearable textiles more and more. The rising numbers of engineering **textile** solutions that have won awards – e. g. The Material Prize by raumprobe (Germany), in 2018 and 2020 – for constructive culture reflect the trend of growing demand on (technical) textiles. Also, the material competition initiated by the Trude and Fritz Fortmann Foundation (2018 and 2020) and other awards, like the initiated new sustainable design award for textile engineering – and machinery by the Walter-Reiners-Foundation of the German VDMA [88] – are underlining this trend. Thus, we can conclude that the **worth of design engineered textiles and the discipline of textile engineering – not only in fashion and sports design – will have a greater impact on transdisciplinary applications**, like medical design and building culture, in the future. The **super 'performance' textiles are more about creating simple everyday solutions for the 'new normal'**. Besides, 'sustainability' is not an option – it's a fact.

'with case studies beside, we can illuminate different education strategies in industrial design engineering and textile engineering in Europe, that you are able to look for that profile, which fits to your pre-educational level and interest'

(M.-E. Wachs, 2019, Glasgow)

To underline the thesis above, the next comparison of the industrial design process and fashion/textile design shows different strategies, and each student's pathway to preparing for and finding the right university, which depends on their pre-design educational experience.

One German case study at the Hochschule Niederrhein is promoting an educational system in 'design engineering textile'[89], which is in line with industrial design engineering: They begin by analysing the process and textiles that revolve around material performance. This is achieved, for example, by using body mapping and textile layering, with the help of a warp knitting process for sports-tech trousers or running shoes. Other European study programmes are taking a more physics-based approach by looking at the process and developing other solutions with alternative design methods – e. g., the Technical University in Dresden and Strathclyde University in Glasgow. Some European textile engineering programmes are using more artistic approaches, according to the university's profile as a university of fine arts. In the past, academic education in fine arts had their primary focus on artistic-based education for design objects, from the 1960s until 1980s. (see Bürdek, B., 2005)

---

[88] Note: VDMA – Verband Deutscher Maschinen- und Anlagenbau, Voice of mechanical engineering – in Germany and Europe link: https://www.vdma.org

[89] See Hochschule Niederrhein, (2020), Studiengang Design Ingenieur – Textil, https://www.hs-niederrhein.de/fileadmin/dateien/FB07/Studium/Curriculum/Modulhandbuch/PO_2017/Modulhandbuch_DI-Textil.pdf. [Accessed on 2020, 04 January]

Nowadays, the changing habits in the way products – objects – are designed, are serving as catalysts for applying a more holistic approach to theoretical methods. This begins with sports engineering courses at the higher educational level (master and PhD design students), and is supported by systems-oriented design methods and theory-based design thinking – with focus on conceptualising.

Generally speaking, the 'playgrounds' founded more on biology, physics and natural sciences, are used for master programmes in sports engineering (see Strathclyde University and Dresden University). These implement a classical and conventional design process: a process, which is based more on material analysis and can be directly measured in economic profits.

Although, different educational approaches related to industry and consumer habits have been demonstrated in Europe, imagine if the user, the designer or the entrepreneur, whoever is using the product, had a greater insight into the design system at an earlier point, rather than having to wait until university – as early as pre-school, for example. Several renowned architects claim that playing with wooden cubes in pre-school during their childhood, paved the way for their future careers. He or she could be viewed as expert from those very first hours onwards, in a (design engineering) playground. Experiencing such simple processes, optimised by the support of a didactic environment, builds a foundation for future learning and exploration. We can see this being embraced in the UK, where teenage pupils can join design engineering courses at school, allowing them to participate in the courses, practice the methods and learn expert terminology in a more playful manner and an earlier age. This indicates that 'design' has greater acceptance in the curriculum in Great Britain, and has a higher didactical worth.

This vision is considering cause and effect for the overall future of our educational systems in Europe: **We need earlier design education (strategies) at primary schools, which may coincide with a change in the terms 'design' and 'industry'.** It shows great advocacy when the director of the Victoria & Albert Museum, in the year 2019, Dr. Tristram Hunt, supports the idea that Prince Albert probably would have dedicated himself to making a big impact on today's design education at school. [90]

The statement 'to develop technically, ergonomically and aesthetically convincing products and innovation studies for tomorrow's working environment' [91] reflects the aim of the Industrial Design Engineering programme at the Technical University of Dresden, Germany.

This shows us that, although design engineering solutions often set out with the same goal, the pathway towards innovative designs and concepts varies across Europe. **We can benefit greatly from comparing, discussing and reflecting together: diversity is the key to our success** reflecting together in Europe: (see chapter 3 and 4).

[90] See Hunt, T. (2019), in: Journal ICON I, Issue October I 2019, Welt am Sonntag, p. 78 f. online available: https://www.welt.de/bin/ICON_ICON_Oktober2019_100dpi-201164330.pdf [Access on 2020, 04 January]

[91] See Dresden, Chair of Industrial Design Engineering. (2020), https://tu-dresden.de/ing/maschinenwesen/imm/td?set_language=en#intro [Access on 2020, 04 January]

6.3.4 Parameters of the education model: Estimating the value of 'artistic engineering designing' processes

6.3.4.1 Materialising your design ideas

'In the last two years it was shown, the demand for high value designed products created across the future European landscape require new educational talents working seamlessly across integrated analogue and digital platforms while responding to evolving cultural needs emerging through new behaviours […] **connected European learning landscape to increase creative diversity***.'* (Wachs M.-E. and Hall, A. 2019) On the one side, higher scholarly educational systems tend towards being design-theory based: that means that PhD programmes place a higher value on 'thinking design' (Rittel) concepts, used to design solutions for highly relevant societal problems today. Complex problem solving is trained by means of concept-based design theory work. On the other side, this concept has to be mediated and the lack of visualising immateriality – caused by theoretical design concepts – needs to be addressed. The following figure shows the profitable result of the 'Fashioning Furniture Future' workshops (see chapter 4), and demonstrates that the 'materialisation' of design concepts in 'design doing': using material rather than being virtual-based, is very helpful within an artistic or practice-based playground. Here, ideas are transferring into material visualized concepts. **The aim is to express, mediate, communicate, and visualize, as well as get people involved in discussions, while being able to evaluate your thoughts by using three-dimensional, material-based sketching.** The following picture (see picture 08, p. 204) shows a case study in which the PhD student is working on visualising and mediating the ongoing process of the illness, dementia. This is expressed through textile and form, as well as colour and surfaces, and represents the method of **'materialising immateriality'** (Wachs, M.-E., 2019) with different qualities. This design method involves different senses, the tangible, tactile one is very beneficial.

Picture 08
PhD Student B's case study shows a visualisation of the development of the illness dementia, in London, 09/2019. During the workshop at Royal College of Art London, the exercise called for 'materialising immateriality' – or 'sketching with textiles' (see M.-E. Wachs, 2020, Bozen).

6.3.4.2 Estimating the Value of the 'Artistic Engineering Design' process while using artistic intelligence alongside Artificial Intelligence

The process of 'Artistic Engineering Design' (Wachs, M.-E., 2020) explores a concept that gives us the opportunity to reflect on the needs of tomorrow's creative processes: It takes the perspective of art, design, engineering and humanities into consideration, while simultaneously providing a framework for evaluating each one of these disciplines. In addition, the societal value of a holistic and humanity-centered education in design engineering, which benefits from fine art's freedom, integrates people around the globe and is helpful for solving problems that emerge, as a result of the global connections between human beings.

Needless to say, holistic and humanity-centered education begins early on. The playgrounds in pre-schools are spaces for exploration and creative expression – a concept that needs to be valued and mediated through all phases of education: 'because everything is language' (Rand, P. 2008), everything is design! To read, to understand and to solve the complex design problems of the post-digital era, we need this kind of valuable design engineering education system, to make the **universities attractive as learning landscapes for everyone and create educational landscapes for every age – lifelong learning landscapes. The aim is to fill in the gap between theory and practice – between thinking and acting – and to create the habitus** which society needs to form through its anthropological view.

This research aims to identify the key characteristics of a valuable culture-based design education, which forms a variety of design languages in education, as well as impacting societal and individual habits. – This post doc thesis based book, written with a more understandable, popular-scientific style, is an offer, to discuss formats of the future, being invited to formulate questions, together. – Textile engineering and 'fashioning'[92] sustainable future solutions, using a holistic approach towards energy-saving and waste-reducing processes (Rissanen, T., 2017), is exemplifying an earlier consciousness about earth's limitations. Yet, we still have to discuss and review this on a regular basis in terms of design engineering education systems across cultures (see Hall, A., 2017).

We need a considerate holistic design education (see chapter 6.3.1) that is not afraid of a discipline's frontiers, but supports true togetherness (see Wachs M.-E. and Hall, A., 2019) and tolerance of 'otherness'– as seen in collaborative pan-European, non-hierarchical learning systems. By thinking and acting together, in design engineering education and expert playgrounds (like the cooperative, interdisciplinary projects have shown), possibilities emerge that also illustrate how this style of **interlinked designing and learning** could be implemented through future platforms of educational playgrounds. These playgrounds will be characterised by philosophical experiences that augment more **'corporeality and impulsiveness' for a holistic approach. The goal is to transfer knowledge as early as possible, in a playful and valuable way – thereby acknowledging the value of time**.

---

[92] Note: 'Fashioning' in this circumstances means the metaphorical sense like 'Fashioning the Future'- creating trend based innovative technical and human related solutions.

6.3.5 Holistic design education will lead to economic and social benefit – the design method of 'materialising immateriality'

The greatest challenge for the global design community and society will be **to design solutions for educational systems**, as well as to impact behaviour and attitudes towards diversity and integration – particularly regarding the global movement of human beings. This book is designed to encourage the Ministers of Education and Economy – and stakeholder, entrepreneurs and others – to place an equal value on arts and design – in relation to the STEM subjects – as it will play an equally significant role on the social impact of smarter sustainable life conditions, supported by the 'management of knowledge' (see chapter 7).

The transition to a holistic design education, linked to earlier interdisciplinary acts, e. g., in pre-school and primary school, also requires more acceptance of the humanities, as we see a shift in design, and cultural habits today. It also demands a certain seriousness regarding in terms of managing and improving the financing worth of education and the educators. By encouraging the Ministers of Education and Economy, they are asked to show courage and emphasize the benefits, as well as to give impulses and facilitate change. However, it takes more than just that to realise, motivate, and act with humility – as has been shown in chapter 6.3.1. For a change to happen, the valuable visions in design engineering and the ideas behind them need to be applied to a system. (see graphic 08, p. 207)

As discussed, there is an emerging need for artistic and emotional intelligence forces that reflect the cultural behaviour within a complex design process and in complex interlinked design education landscapes.[93] **As we move forward, the behaviourism related to sustainable materials and sustainable design education systems – which is formed before university – will require the basic skills provided by artistic education. These skills will be needed, in order to be able to correctly estimate the ethical values of advanced societies in the future.** In addition, they will be essential for managing complex design issues in the post-digital era.

So, how do we move forward? 'Integration Design' and 'Integrated Cultural Intelligence' (ICI) are both methods that require cultural empathy and a new cultural intelligence. They should be practiced as the implementation of two very significant developments are beginning to take shape: fighting for sustainable solutions all over the world and integrating beneficial systems of digitalisation as part of the fourth industrial revolution. The current challenge is to combine optimised hybrid (virtually connected) classrooms AND analogue experiences in materialising design concepts together.

[93] See Wachs, M.-E. and Hall, A. (2019), European Driving Range – innovative landscapes for a tangible, non-hierarchical learning space within a material and immaterial togetherness, (EPDE, Glasgow, Scotland); ibid: Gulden, T. and Støren Wigum, K., H. W. (2019), Keeping up with Compexity, (EPDE, Glasgow), Scotland, p. 404-409; ibid: Wachs, M.-E. (2018), Driver for sustainable (industrial) design culture – the >design shift<, (EPDE, London), p. 394-399.

QR Code 07
EPDE 2021, No. 104:
see the movie
of presentation
at this website

QR Code 08
PEM – 'Programm of
European Mentoring' by
Marina-Elena Wachs,
see more about the project
and mentees' inspiring
design engineering
solutions and European
interactive insights.

Considerate holistic design engineering education implements cross-cultural connectivity in the digital worlds of tomorrow, and demonstrates **a multi-sensory, emotional and intellectual understanding of complex industrial systems and design education systems. Both of these systems are profiting from this new value in the age of knowledge**. And, although this age of knowledge is characterised by combining systems operated by artificial intelligence, it is the emotional intelligence of future complex problem solvers that is more important. The cross-cultural experiences of 'materialising immateriality' in design has shown that complex problem solvers as design engineers and textile engineers will need this emotional intelligence (as 'homo hapticus', see Grundwald, M.) to address the issues we will face as a society in a changing world, with changing habits. For developing and promoting integrative cultural intelligence, an interlinking of the systems and trigger points of advanced knowledge management will help us solve the problems that are expected to evolve in this interconnected, fast-paced world. **Further exploration of considerate holistic design education will help** us realise that the indirect financial profit that investing in education brings, is several times higher than direct economic profits in the short term. Investigation in holistical education landscapes is the key.

Graphic 08
Key factors for future
design engineering education,
Marina-E. Wachs, 2021

'If the next experts of designer
and engineers are mentors
for children[…] the loop will be
closed and guarantees sustainable design and engineering
education, please see: the
graphic as circle of 'key factors
for future design engineering
education' […]' M.-E. Wachs,
EPDE 2021, paper No. 104.

Please see also conference
paper as presentation:
QR Code 07 and QR Code 08
about European mentoring
engagement.

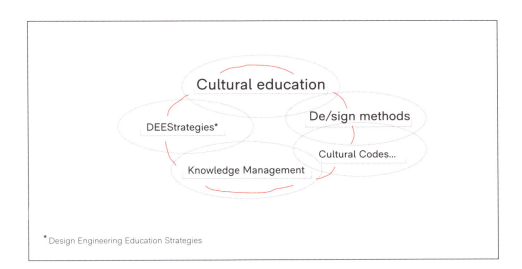

IMAGES 03

International Guest Statements

With graceful thank to all my international cooperation partners, friends and family, in the following you will find statements by invited experts around all creative disciplines, across cultures, across generations for creating our future together, to a turn for a better. Molto grazie, sinceras gracias, vielen Dank, hjertelig tak, uppriktigt tack, tusen takk, merci mille fois.

Ulrike Brandi DE
Moritz Schwarz CH
Karl Borromäus Murr DE
Anne Louise Bang DK
Marina-Elena Wachs DE
Ashley Hall UK
Max-Luca Wachs DE
Adrian Franken DE
Louis Reigniez FR
Theresa Scholl DE
Charlotte Sjödell SE
Olaf Redlin DE
Björn Stichnothe DE
Jennifer Wetzel DE
Wenche Lyche NO
Giulia D'Aleo IT
Valentin Wachs DE
Sibylle Klose DE
Maximilian Krummen DE

Ulrike Brandi

'natural light is my
adored teacher'

Over the last three decades I worked as a lighting designer as part of the global design community. During this time, I have witnessed and experienced significant challenges to our profession, responsive to and directly related to the social, cultural and political changes that roil our world.

While studying with Professor Dieter Rams at the Hochschule für bildende Künste in Hamburg, his teachings regarding industrial design enhanced ideas which I had been developing as a young adult. In his classes, we designed quiet but impactful objects which serve the individual. This was called 'zeitlos' or 'timeless' design. Then, students were protesting against capitalism, challenging profits over people, celebrating squatters and marching against nuclear power. As a response, the design faculty at Hochschule für bildende Künste as a former Kunstgewerbeschule, was focused on design for mass production, rather than the making of luxurious products. But ironically former students entered craft professions as artisans, furniture designers and other 'niche' pursuits, as industrial employment at the time was scarce. When I began to work as a lighting designer, this was still a very seldom and specialized practice. To a large degree it still is today. I had never worked in a lighting studio, so I had to develop my own ideas and methods. It meant confronting the complexities of articulating concepts to others, producing those designs with precision and integrity, and working to see them integrated into whole buildings and places for people. Drawing by hand and physical modelling became my valuable tools.

And as much as I respect the design and construction processes in place today, it is my firm belief that the digitising of all we do has subordinated our physical methodologies with the result that we are losing the very intuitive and some would say, sensuous nuances which produces work of value. If young people are not trained to exercise these basic brain/hand functions, I fear a big loss for our profession. Not only lighting designers, also our clients started to prioritise the detailed calculations instead of the general view. They trusted my competence as a professional when I drew red arrows to show our concepts in a simple section. Nowadays clients demand for visualisations, produced with a huge effort of time and technology to be convinced of the concept. Is this a loss of self-confidence? I am far away from proposing to put the clocks back, I am searching for a fruitful combination of artistic, intuitive and analog design methods together with systematic, disciplined and complex digital methods. To combine the knowledge of multidisciplinary teams, to perform in complex communication acts, to reflect these and to care for the attention to the sense and the meaning of what we do, is the ideal I am searching for in each design process.

These two images are typical for our approach in methods and ethics:

1. The photo of our lighting design for the Elbphilharmonie in Hamburg shows the relation between visitor and natural environment, the view out of the window and the calculated filtering and distribution of natural light and heat.

2. The photo of the Central Station of Rotterdam shows this point of traffic. A Rotterdam citizen praised it as the new living room for the City of Rotterdam. Lighting design is for the people!

Ulrike Brandi
is the director of Ulrike Brandi Licht GmbH since 1986 and the Brandi Institute for Light and Design. Her office planned around 1.000 projects around the world, including the Elbphilharmonie Hamburg, Day, Light and Night exhibition in the Istanbul Modern Museum, Munich Airport T2 and satellite terminal, Pudong International Airport T2 in Shanghai, Amsterdam Central Station, Rotterdam Central Station, the Rotterdam Masterplan and the Royal Academy of Music in London. She published various books and teaches at international Universities.

Contact
www.brandi-institute.com
https://www.ulrike-brandi.de/en/profil/

A
Elbphilharmonie-Hamburg Plaza
'The photo of our lighting design for the Elbphilharmonie in Hamburg shows the relation between visitor and natural environment, the view out of the window and the calculated filtering and distribution of natural light and heat.'
Fotograf: Jörn Hustedt
für Ulrike Brandi Licht

B
Central Station of Rotterdam
'The photo of the Central Station of Rotterdam shows this point of traffic. A Rotterdam citizen praised it as the new living room for the City of Rotterdam. Lighting design is for the people!'
Fotografin: Ulrike Brandi Licht

Ulrike Brandi

Moritz Schwarz

A

'the architect withers into a form-specialist who no longer holds the reins in the organizational chart'

An architect's challenge: bridging the creative freedom with the forces of optimization like BIM

'As the author of the draft and the head of the further planning and execution of a building, the architect contributes to shaping and designing the living space of the people. He takes responsibility for a design and execution that meet the requirements of the client, the environment and the public.'[1]

This guiding principle ought to reflect the ideal of the architect as a generalist who, striving for the best possible design, contributes his knowledge and set of skills with every arising opportunity. In my professional everyday life, however, I see a fragmentation of competencies, because the construction tasks have become complex – to such an extent that not only one discipline, i.e. the architect, can oversee all areas. The complexity of the statics, building services, building envelope, building physics and traffic planning exceeds my general knowledge as an architect. For this reason, universities have increasingly shifted their education on the design, other subject areas are only taught peripherally for a rough basic understanding. The architect withers into a form-specialist who no longer holds the reins in the organizational chart. He is no longer authorized to issue instructions on behalf of the client. For this reason, service-provider offer general planning and overall coordination on behalf of the architect. Do architects need to reinvent themselves, and can BIM help them? My academic career has led me to the precise detailing of the ETH Zurich via the free-thinking College of Art in Edinburgh. I was able to get to know the computer-aided design methods early on, for example during an internship at UN Studio Architects (Amsterdam). These deliberately sought the dissolution of the classic order of vertical wall and horizontal ceiling. The resulting spatial free forms could be implemented on a small scale with the help of CNC and later 3D-plotters, but were hardly feasible for formwork plans such as the Mercedes-Benz Museum (Stuttgart) or the Phaeno Museum (Wolfsburg) by Zaha Hadid Architects. The hoped-for geometric liberation, supported by free form generating programs such as Rhinoceros, was hardly used in practice. The aspect of programming, especially the parameterization, for example the linking the height of the geometric components to the room height, laid an important foundation for BIM. BIM, Building Information Modeling, describes a *'cooperative working methodology with which, on the basis of digital models, the information and data relevant to a buildings life cycle are consistently recorded, managed and exchanged in transparent communication between those involved or passed on for further processing.'*[2] The core element of BIM is always a virtual 3D model, the digital building model, which is enriched with information from various specialist disciplines.
This model is a representation of the planned project. Beside the geometry of the construction, alphanumeric information can also be attached, for example the surface quality, the target price per element, the time of installation or the time of revision for facility management. Despite the risk of an inadequate level of detail, there are almost no limits to the information density. With the appropriate effort, the virtual building model can be made available to everyone involved in the project, across all phases and disciplines.

BIM found its way into practice particularly in the Anglo-Saxon region; in Great Britain, for example, over half of all planners were already working according to the BIM method in 2014[3]. It turns out that this adaptation was sometimes carried out too quickly: instead of a uniform role and process definition, the focus was heavily on the technical implementation. Germany has chosen a more moderate introduction of the BIM methodology, but the *Stufenplan Digitales Planen und Bauen*[4] also requires that all infrastructural buildings must be carried out with BIM from 2020 onwards. Such requirements have forced German planners to deal with BIM at an early stage, so that the BIM method has become part of everyday work in many offices. Although the architects could choose their own way of working in the interests of freedom of method, the pressure from customers is growing. The economic interest in a faster and more cost-effective planning with the help of BIM inevitably leads to a standardization of the components, in which a building is copy-pasted from a library of BIM objects. In my opinion, this is a contradiction to the individual, location-specific spatial solution which makes for good architecture.

However, this development will produce a job profile with a strong focus on data management, which in the worst case will not have an understanding of the spatial setting of the usage-related relationships. In my day-to-day work, I experience the decreasing time invested in developing, trying out and sometimes discarding of design ideas, but rather quickly taking the step to 'implementation' of a quick solution into the digital building model. A lot of energy is wasted with the correct programming, the readable display or the correct export settings. **Time and resources can only be freed up for increasing the architectural quality of the design if the digital tools recede in their importance and the various models are handled as intuitively and smoothly as possible.** The sketch, quickly and clearly put on paper, is ahead of any virtual model. This craft of depicting a spatial situation in drawings must remain a core competence of the architect. At the same time, he must be able to convey, convince and arbitrate what requires communication skills and the ability to work in a team.

I therefore see it as the architect's duty to proactively take on this coordinative role so that he can ensure the realization of his initial design idea. In my opinion, architects will have to strengthen their social and technological skills. With the right basic knowledge and a natural use of both, analog and digital tools, the architect can again consolidate his original role as overall manager.

I would like BIM to be seen as an addition to our competencies solely. Only with a holistic repertoire of our skills, entirely in the spirit of the generalist, can the quality of the architecture be improved and continue to be guaranteed.

Moritz Schwarz
Born in Karlsruhe (D), graduated in Architecture at the RWTH Aachen (D), after gaining work-experience in international offices like UN Studio (NL) or GMP Architekten (D), after realizing several bigscale projects in Zürich, he analyzed the way of collaboration in heterogenic teams during his MA Thesis at the ETH Zürich. Focusing on the use of BIM (Building Information Modeling), he helped to shape an understanding of this new methodology for Swiss architects and specialists. He lives and works in Zürich (CH).

Contact
Moritz Schwarz, Architect at
PENZISBETTINI. Architekten GmbH, Zürich
Alfred-Escher-Strasse 23
8002 Zürich

www.moritzschwarz.com
info@moritzschwarz.com

A
Screenshot during an internal design-workshop (with Microsoft Teams).
Screen by Moritz Schwarz, with friendly authorization of all members, 2021.

1 see Schweizerischer Ingenieur- und Architektenverein (Hrsg.), SIA 102 – Ordnung für Leistungen und Honorare der Architektinnen und Architekten, Zürich, 2013, p. 46. **2** see Bundesministerium für Verkehr und digitale Infrastruktur (Hrsg.), Stufenplan Digitales Planen und Bauen, Berlin, 2015, https://www.bmvi.de/SharedDocs/DE/Publikationen/DG/stufenplan-digital-es-bauen.pdf [Access on 2017-30 – October]. **3** see National Building Specification (Hrsg.), NBS National BIM Report 2014, Newcastle Upon Tyne, 2014, Adresse: https://www.thenbs.com/knowledge/nbs-national-bim-report-2014 [Access on 2018-12 – January] **4** see Bundesministerium für Verkehr und digitale Infrastruktur (Hrsg.), Stufenplan Digitales Planen und Bauen, Berlin, 2015, Adresse: https://www.bmvi.de/SharedDocs/DE/Publikationen/DG/stufenplan-digitales-bauen.pdf?__blob=publicationFile [Access on 2017-30 – October].

Karl Borromäus Murr

Advocating Responsive Design

This statement argues in support of a responsive design approach which is in line with the concept of 'slow fashion' providing an alternative to the unrelenting and destructive growth logic of 'fast fashion'. Sustainable fashion accepts ethical responsibility in its design by reacting sensitively to the environment when creating products. Along with Mathilda Tham, Kate Fletcher places the local dimension of design requirements at the centre of her considerations. Instead of the dialectic of centre and periphery, which serves the global West and places Africa, for example, at the edge of the world, Fletcher and Tham plead for a significant increase in the value placed on locality, which, in contrast to centralism and territorialism, emphasises a decentralised world with many equally important scenes of action. The commitment to locality – Fletcher and Tham speak of 'localism' – also includes work with geographically close local and regional value chains. In addition, the concept of 'localism' stands for the appreciation of unique local social forms, traditions, knowledge cultures and trade practices – cultural particularities which are to be honoured not in the sense of social-romantic nostalgia, but rather in the sense of societal empowerment. The reference to 'empowerment' points to the socio-political alignment of the concept of locality, which aims for a new distribution of power that will, in turn, strengthen and stabilise the responsibility of local communities for developing their own decision-making capabilities and options for action. Localism in design hence runs counter to a globalised economy with its homogenised and standardised products. In contrast to the accelerated logic of economic scale effects, localism emphasises a deceleration of products and consumption – a localism that hence reassesses the level of demand and favours the more sustainable quality of products over mere quantity.

Dr. phil. Karl Borromäus Murr
Director of the State Textile and Industry Museum Augsburg
Karl Borromäus Murr was educated in History, Philosophy and Ethnology at the universities of Munich, Oxford (St Edmund Hall), Eichstätt-Ingolstadt and Cambridge (Harvard University).
In 2005 he received his PhD in History from the Ludwig-Maximilians-University Munich (LMU), Germany. In 2009 he was appointed director of the State Textile and Industry Museum Augsburg. Since 2015 he is a board member and since 2019 the chairman of the European Museum Academy. Since 2005 he has been teaching history, ethnology and museology at the LMU and the University of Augsburg. He has published widely on history, art history, ethnology and museum studies.

Contact
karl.murr@timbayern.de.
https://www.timbayern.de/ausstellung/dauerausstellung/

Anne Louise Bang

'learning from the past, living in the present, designing for the future'

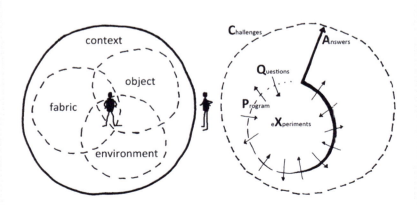

One of the most interesting aspects of being a teacher and researcher in textile design is the constant need to challenge and change the field. The world as we know it is rapidly changing. We experience a severe global environmental crisis, and digitalization encourages us to replace and adjust known processes. Environmental, technological and societal challenges have a huge impact on our educational systems. We need to educate for lifelong learning addressing rapid changes of a fragile and unpredictable world. While still relevant, it has for a while not been sufficient to educate candidates that can manage all processes from idea to product to market within the field of textile design.

Textile designers must be skilled experts trained to work in cross-disciplinary teams. However important, it is not enough to teach basic skills such as weaving, printing, knitting, embroidery, etc. Sustainability is no longer a parameter among other design parameters. It is now mandatory to include it in all aspects of textile design. A textile designer must be able to design and work with the material as part of an object in an environment while constantly considering the overall context. At one and the same time, I encourage students to train as experts as well as team players. I emphasize the need to choose an expert area in the broad field of the discipline and to do it in a way that allows the students to work beyond a narrow understanding of disciplinary tools and techniques. Thus, I provide the students with skills and expertise that meet the demands of the industry and the SMEs, and, in parallel, I prepare them to work independently, innovatively, entrepreneurially, cross-disciplinarily and with a socially-minded and responsible mindset.

If textile design is as broad as that, what is then the characteristics of the field? I strongly believe that experiments and prototyping play a major role in textile design and that crafting goes hand-in-hand with thinking. Therefore, it is crucial that textile designers possess the competencies to choose relevant tools and techniques depending on the context and application. They must be able to define the problem or the underlying challenge of the brief and sometimes do this in opposition to the industrial or societal wishes and demands. Core skills such as construction techniques, surface design, aesthetics and co-design supported with knowledge about materials, environmental issues, societal challenges, product attachment, use patterns and circular value chains should be at the core of textile design education these years. The ability to learn from the past, live in the present and design for the future must be core elements.

Anne Louise Bang
Senior Associate Professor PhD
Dr. Anne Louise Bang is a Senior Associate Professor in Design & Sustainability at the Centre for Creative Industries & Professions at VIA University College in Denmark. Bang has a background as a textile designer and weaver. Her research focuses on experiments and development of practice with a special attention to issues of sustainability. Her current work builds on her PhD studies, where she developed tools suited for a dialogue about sensuous and experiential aspects of textiles.

Contact
anlb@via.dk
https://en.via.dk
www.via.dk/uddannelser/tekstildesign-haandvaerk-og-formidling

A
A picture by 2020, view inside the print lab VIA University College, DK, by Inger Marie Ladekarl

B
My understanding of applied textiles and the central role of experiments in design.
Design graphic A. L. Bang, 2021

Marina-Elena Wachs

Designing 'Cultural Resources'

Prof. Dr. phil., Dipl.-Des. Marina-Elena Wachs
is master tailor and industrial-designer and professor for design theory. 2007 she graduated at Braunschweig university of art with her doctoral thesis ('material mind – new materials in design, art and architecture', https://verlagdrkovac.de/978-3-8300-3292-2.htm). Since 2010 she held a professorship in theory of design at Hochschule Niederrhein. Marina's present research focussing on > interdisciplinary projects in sustainable design<, >textile engineering in cross cultural learning landscapes< and >sketching the future: design and children<. Note: In the year 2020 Marina got a call for Professorship for 'Designo' (Italy)/Zeichnen (German)' at the faculty of educational sciences, at Free University Bozen, she rejected the offer.

More research information
https://www.researchgate.net/profile/Marina-Elena-Wachs
https://www.linkedin.com/in/marina-elena-wachs-551662a5/
Next publication 2022, which is developing the research subject in this book further more: 'The art of languages for good design – A didactic approach to art + design in pre-school education + Design engineering mentoring culture', Wachs Marina-Elena.

'A pan European transformation to bridge between tangible experience and virtual ideating spaces', Marina-E. Wachs together with Theresa Scholl and Giulia D'Aleo in preparation for 2022. More information to PEM – Programme of European Mentoring, see website (see QR Code 08).

Contact
info@marinawachs.de
www.marinawachs.de

Sustainable design engineering – and nothing less. This begins early on, at preschool age, when designing, creating, working, to gestalt or shape things, imagining, constructing, and building with hands, mind and body, takes place. Also, it means collaborative, cross-cultural, synesthetic creating, with a holistic approach: involving art, music, poetry, and experiences with natural materials, while reflecting on historical and present-day narratives. But, to be able to do this, you need a solid foundation... one that allows for playful designing with self-confidence in stress-free learning spaces. (see pictures at p. 222 f).

It is my vision, that more designers and design entrepreneurs will dedicate themselves to more in-depth design mentoring – as part of university curricula, within enterprises and in social circles, like pre-schools. Let us train the next student generation to become mentors for children as well as industry managers, while they still have the opportunity to learn from the older generation of 'golden' mentors'. We must encourage new students to take the initiative to interact with these mentors and integrate them into design and cultural education – to cultivate a 'culture of questioning' is our goal.

After more than 30 years of creating as designer and more than 20 years of working as an educator – of which more than 10 years as professor for design theory – I have collected so many thoughts, activities, experiences, operations, projects, events, journeys, reflections, workshops, as well as interactions with friends, family and mentors. All of these have sparked questions about design engineering over the course of history, as well as in the present time and the future – and particularly in the field of design engineering education. With my German perspective – especially in comparison to the Northern Countries, such as Sweden, Denmark and Great Britain – I have to thank so many creatives and thinkers, who have collaborated together with me: Ultimately, I can say: It is not a question of struggling with and rethinking design objects, concepts and industrial processes. Instead, it is the human factor that inspires and enables creation; by examining the problems + tasks presented to us by our planet; by solving these issues together in cross-generational + cross-national + cross-discipline + cross cultural collaborations; by going beyond cultural behaviour and obligatory inherited behavioural patterns.

The present and urgent need for 'Circular Process Management' is represented by the 'cycle of key factors needed for sustainable education development' (see graphic 08, p. 207, by Wachs, M.-E., 2021). This circular process is a beneficial tool for education systems beyond schools and universities. It is also beneficial for self-education and coaching management in enterprises and industries, as well as for society as a whole. In the future, cultural education needs to embrace the HUMANITIES – sometimes derogatorily referred to as 'the little courses' in Germany – to the same degree as the STEM subjects. Please invest more in these 'little courses' – in the humanities and mentoring in human 'personal intelligence concepts'. After all, these will train the emotional and cultural intelligences that will be essential as we move into the post-digital age, in which the motivation to act sustainably has begun to shift.

Cultural education + knowledge management are the biggest resources for humans + the planet – let us take a stand for the resources that are most important for developing future design engineering solutions. This can be achieved by investing in sustainable cultural education – by teaching future experts, children and other 'knowledgeable creators'.

My graceful thank to all who participated in, and supported the creation of this book – to all who designed, reflected and discussed with me, making it such a valuable book to read – for gaining new sustainable and holistic insights and discoveries – with joy.

Marina-Elena Wachs during some basic 'design doings' during the first classes of industrial design study within the laboratories 1995/96 at Hochschule für Bildende Künste (HBK) Braunschweig, pic: M.-E. Wachs.

A
Painting like the renaissance artists, still life with natural based painting materials – different technics have been trained.

B
Hands on design – experiences with stone, gipsum, Balzer wood, metal and synthetics. Exercises to 'form and aesthetics'.

C
Laboratory of gypsum – elementary course, hands on design, M.-E. Wachs, 1996.

D
Marina-Elena is sketching within the Augmented Reality, 2021, in VRHQ lab Hamburg, Germany, thank to Roland Greule!
https://www.haw-hamburg.de/ftzdigitalreality/

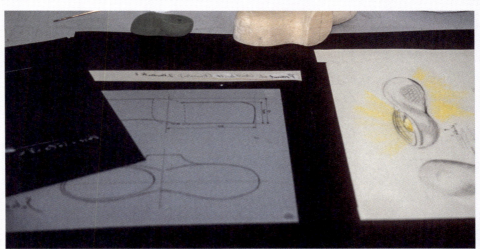
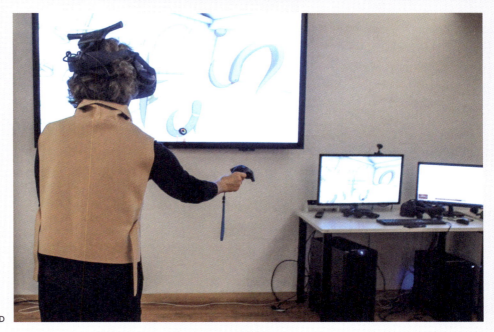

Marina-Elena Wachs

Ashley Hall

'at the heart of this was an effort to train failure resilience and to make failure a natural and desired part of the innovation process'

Marina-Elena Wachs:
'What do we need to improve the future of design engineering education?

Ashley Hall as Professor of Design Innovation answered:

I believe we have design futures, not a single future. We have a choice and need to find better ways to arrive at those better futures. There are many ways of doing this and in design education we have a wide range of tools and methods for designing futures. These mainly fall into two groups; the first is anticipatory (Fuller, 1992), we project an imagined future then design a solution and wait for it to happen. The second is to envisage the futures we want to arrive at and proactively design for steering towards them. The first set is useful more for applied design and the second more strategic and involves working outside of traditional design relationships with industry.

This is an even more pressing issue now that we understand that design needs to develop a more proactive model for us to tackle the big global issues that are coming our way in the future. Proactive or prospective design (Galdon, Hall, Ferrarello, 2019) helps us extrapolate preferred future destinations and design our trajectory from today into the future.

You could say that all designing happens in the future. John Chris Jones give a good description of the relationship between design and the sciences in relation to the future:
> 'Both artists and scientists operate on the physical world as it exists in the present (whether it is real or symbolic), while mathematicians operate on abstract relationships that are independent of historical time. Designers, on the other hand, are forever bound to treat as real that which exists only in an imagined future and have to specify ways in which the foreseen thing can be made to exist.' (Chris Jones, 1992)

While Ranulph Glanville captures the unique quality of designing:
> 'I propose that research in design should forge a new type of knowledge, *knowledge for* [Future Transformation].' (Glanville, 2005).

It's this *knowledge for* transforming the future as opposed to *knowledge of* the present that can steer us towards a more significant role for design engineering by bridging the society-technology gap by specifying how these new foreseen things can come into being. How can we steer towards better futures and who decides what better looks like, who is included and excluded in the conversation?

Early in 2020 we experienced the first global pandemic played out in real time. It's clear in this new unimagined world that design was waiting to respond. We have seen some impressive products and service design solutions tackling the issues, but the real issue is that we were waiting. We were waiting in the traditional design mode of waiting for a problem to arrive, waiting for the design brief, and we now know that we can't wait any more. We have hit the fast-forward button for design and learnt that strategic avoidance is better that tactical responses to a crisis. The future of design and engineering design have to question our practices and assumptions about the relationship between designing and the future and how we educate the next generation of designers.

We decided to tackle this issue via a design for safety grand challenge at the Royal College of Art's School of Design with 380 postgraduate design students and over 20 academic staff and invited experts with a major project over a four-week period sponsored by Logitech. The students worked in groups and were spread around the world in over 50 countries and used online collaborative tools to look at this challenge leading towards a new model for design. We looking at how design could tackle design leadership, design futures, care, health, design for resilience, design for truth and next generation interactions to move beyond waiting to act. We estimated that we dedicated 32 years or 62.000 hours of creative capital designing new responses to emerging and projected future global issues.

A few years ago, we started noticing that our Innovation Design Engineering postgraduate students were becoming increasingly risk-averse. The pressure of increasing fees and having to maintain good average grades for scholarships alongside disciplinary norms to avoid risk and keep to tried and tested formulae was having an impact. To address this, we decided to run a failure project where students would be graded in terms of their ability to fail, the better the failure the better the grade (Hall et al, 2016). At the heart of this was an effort to train failure resilience and to make failure a natural and desired part of the innovation process. Something that indicates work that is finding the cutting edge rather than avoiding getting it wrong. In education we should always reward brave failures rather than modest successes as that is what is at heart of great design led innovations. Edison said 'I have never failed, I found 10.000 ways that didn't work' and this captures the central place of finding edges in successful design, whether the edge is technology performance, public acceptance, economic or sustainable.

A large proportion of the global population live it cities with millions of people and hundreds of millions of products. Many of these products are directly related to our own safety and wellbeing yet we have few if any education programme dedicated to designing for safety. Why is this? We know that engineering considers safety factors and has effective practices for safety yet design is also a safety critical activity and cannot solely rely on engineering for safety. This is especially true as we move into more experiential types of design and start to design with increasingly complex systems including AI that go beyond human comprehension. We need to begin thinking about new ethical and safer relationships with this technology.

At the heart of this is designing resilience, the capability of flexing to cope with expected and unexpected future changes. A big part of this involves designing safer failure spaces and understanding more about how we educate designers to develop better methods for sophisticated ecosystem-smart solutions to wicked problems.

Across design education we need more than ever to rebalance the society-technology gap through a better understanding of the relationship between designing safety and failure to build resilience against an increasing set of survival challenges for ourselves and our ecosystem. On our design products MA programme, we are looking at a new educational ethos to support designing for dematerialisation, subtraction, circularity, decolonisation, delimiting products and multi-species design to start asking the serious questions that will drive future design practice over the next decades and support the education of this new generation of designers. We know what we need to do to improve the future of design education, it's a new model for design.

Prof. Ashley Hall
Prof. Ashley Hall PhD, RCA London, UK
Ashley is Professor of Design Innovation in the School of Design at the Royal College of Art in London. With a MA from the RCA and PhD from the University of Technology in Sydney he has a background in design practice, teaching and research. Hall leads post-graduate research for the School of Design and the MRes in Healthcare Design. Ashley researches in innovation methods, design thinking, design for safety, experimental design, design pedagogy, globalisation design and cultural transfer.

Contact
https://www.rca.ac.uk/more/staff/ashley-hall/

References
Fuller, B. (1992). Cosmography. USA: Hungry Minds Galdon, F., Hall, A., (2019) Prospective design: A future-led mixed-methodology to mitigate unintended consequences, International Association of Societies of Design Research, Manchester, UK Chris Jones, J., (1992). Design methods. New York: Van Nostrand Reinhold. Glanville, Ranulph. (2005). 'Design Prepositions.' In: The Unthinkable Doctorate, 1-9. Brussels, UK and Australia: Cyberethics Research, American Society of Cybernetics. Hall, A., Bahk, Y., Gordon, L., Wright, J., (2016), The Elastic Octopus: A Catalogue of Failures for Disrupting Design Education, Engineering and Product Design Education Conference, Aalborg, Denmark, September 2016.

A
Design Futures Landscape,
Ashley Hall, 2021.

B
Graphic Design Futures
Methods Landscape,
Ashley Hall, 2021.

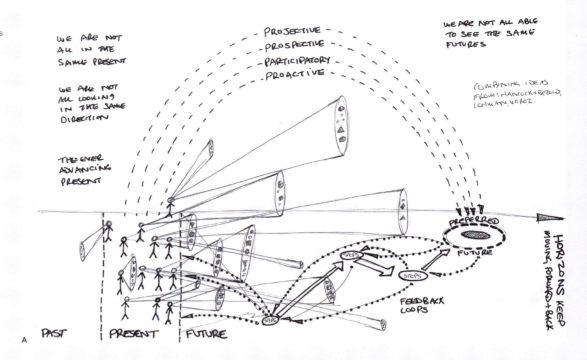

Max-Luca Wachs + Adrian Franken

'humans are going from SSD to RAM'

Form follows // enables // deepens function

As software rendered most of the haptic items in the last 15 years meaningless, humans became androids without realizing it. Now, AI is on the horizon to take over what humans are currently doing in software. Creating things becomes less haptic and practice-driven, as in mechanical realization, and more and more open for creative ideas of functions. In product design, products are turning into services, as the user experience is moving more and more into the focus of product innovation.

Automation like Generative Design eliminates tasks of engineers and designers alike, enabling AI to execute artisanal and solution-finding processes. Depending on the field and direction of design, the optimal form mechanically might even be valuable as the optimal design, in which case the engineer becomes the designer, which is where AI already does its first steps towards executing most of the work, while humans merely have to control the inputs and analyze the outputs (example: AI-designed drone frames, where forces are put in and all that is left to do is choose a favorite form of the generated possibilities). Hereby, especially mechanical engineering has the potential of becoming hugely automated.

With the increasing speed of progress, design and engineering are fused further together, if not already before, as the mechanical realization of products becomes easier to adapt to designs. Functional thinking is not only allowed to engineers, as well as creative thinking is no longer the responsibility of only the designers.

It is increasingly important for designers to take social developments into account and implement core topics like sustainability, social design and new ways of living into one development process. Especially sustainability is relevant for both designers and engineers, as the global industry is coming closer to the limits of available resources, and has to be integrated from an initial idea all the way up to global supply chains, to not further support a system doomed to collapse if continued. Given this development, as the steps of functional dimensioning and finding a form for realization are getting closer and closer together, designers are getting closer to becoming engineers with the need to learn how to validate if their work can function, and engineers are getting closer to form-deciding positions, with the need to understand design language and its relationship with mechanical function.

Max-Luca Wachs
Max-Luca Wachs, currently master's student of International Business Engineering at the Hamburg University of Technology, 2021 study at Politecnico di Milano; Bachelor of Engineering in Business Eng./Industry at the Leuphana University Lüneburg. Minor experience in the Automotive industry, plant planning and entrepreneurial endeavours.

Contact
maxwachsdbs@gmail.com

Adrian Franken
Bachelor of Arts in Industrial Design at the University of Applied Sciences Potsdam
Experiences in the Consumer Electronics, Furniture and Automotive industries.

Contact
info@adrianfranken.de

Louis Reigniez

A

'... our eco-systems are infected with micro-plastics! – we need to act.'

We live in a world where today's decisions will have an important impact on a fruitful and clean future. The world's environments are changing fast and in unpredictable ways, the chances of us having a stable climate are slim if the right decisions are not made in the foreseeable future. Our eco-systems are suffering from the pollution of our daily consumption. Today, we dump more than 15 tonnes of plastic waste into the oceans every minute. In 2030, that number will go up ten fold. Most plastics don't float and what we see on beaches is just the tip of the iceberg, our eco-systems are infected with micro-plastics. We have no idea yet what the impacts of these plastics are on our ecosystems. We must find a way to turn off the plastic waste tap, a short term solution to a long term problem. Our solution is definitely not the best one for the future but it's one that will have an impact on the problem today and tomorrow this type of solution will develop into something more harmonious with nature's needs.
The ideal solution is not to throw away plastics and develop a circular economy, but the problem of plastics is happening today and we need to act.

We manufacture a material made up of plastics that are not well recycled and are often buried, incinerated or exported to third world countries. These plastics have incredible and distinctive mechanical properties that can be useful in many sectors, these should not have just one lifecycle. Our transformation process is fairly straight forward, we identify, we clean and heat plastic into different sheet sizes to then be used for all sorts of purposes, such as furniture design, interior design, product design etc. These sheets can be mass produced and can have a real impact on the plastic waste problem.

We have to make a material that is not only easy to use, but also attractive. We want to make sustainable products look sexy, otherwise people won't be inclined to use this type of material.

Here are some of the colors that we are testing made from plastic packaging found in restaurants. This plastic has amazing properties because it is very easy to work with, does not break when subjected to extreme torsion and has a marble like texture.

Louis Reigniez
I'm Louis Reigniez, the founder of Ocean One. I'm 28 years old and I live in the South West of France in a coastal town near Bordeaux. I have studied and worked in the Banking industry for 7 years in the U.K. and I couldn't just watch the oceans being rekt. So I decided to make my way to the center of the environmental stage, I want to be part of the people who fight for our future.

Reference
Louis Reigniez, No 82, page 36 ff:
http://www.ville-lege-capferret.fr/wp-content/uploads/2021/07/LCF-Presquile-82_WEB.pdf
http://www.ville-lege-capferret.fr/actualites-et-agenda/publications/

Contact
Louis Reigniez
Foundateur
Ocean One
+ 33 7 49 01 60 91
contact@ocean-one.fr
ocean-one.fr

A
Rawmaterial of recycled ocean plastic for new products for the bathroom (soap dish) and plate as cape lifter, design and construction by Louis Reigniez.
Pic: M.-E. Wachs, 2021

Theresa Scholl

A

'as a designer you don't define yourself with the tools you use, you rather define yourself through your own thoughts, ideas and visions. The future in design is not the amount of digital services, advanced technologies or innovative materials, the future in design is the vision YOU have in mind and share with others of how the world would look like tomorrow'

Studying design engineering at a German University means being in a very forward-thinking space. University is heaven for a designer. Everyone is open-minded for innovation and thinking out of the box. Students have the opportunity and are even encouraged to work interdisciplinary. One can explore new materials, try various production methods and always has access to the newest software. Working close to research means being in touch with future scenarios and creating products, services and interfaces which aim to solve tomorrow's problems and needs.

Leaving university and this bubble where production cost, sales figures and legal requirements didn't really count that much and starting the first job in industry, the world suddenly looked quite different. Working in a modern start-up culture, for established brands or mega-companies, one doesn't have the excess to all technologies or the possibilities one had at university. But no need to worry – as a designer you don't define yourself with the tools you use, you rather define yourself through your own thoughts, ideas and visions. And while maybe having lost some tools, modern methods or technologies, the learning from university last for a lifetime. These learning combined with keeping the fearless student's mind of not knowing but trying everything will help you to be successful in your job.

Thus, while entering the industry as a young design talent, you still want to keep up the willingness to learn new things. It is worth and important to respect the corporate culture and understand well-established techniques but at the same time always challenge the traditional thinking. Companies will win by complementing existing and established knowledge with fresh perspectives of young design talents. Especially in the field of product development, it is key to combine the deep understanding of experienced (but sometimes entrenched) colleagues with the fresh (and maybe naïve) view of the newbies to create visionary products.

The future in design is not the amount of digital services, advanced technologies or innovative materials, the future in design is the vision YOU have in mind and share with others of how the world would look like tomorrow.

A
Three generations of industrial designer and design engineers, collaborating and designing with textile and light for a sensible usage with ressources; from left to right: Ulrike Brandi, Theresa Scholl, Marina-Elena Wachs, pic: M.-E. Wachs, 2018.

Theresa Scholl
Color & Trim Designer Volkswagen Nutzfahrzeuge, Theresa Scholl is a bespoken tailor and textile designer with a Masters degree from Hochschule Niederrhein. During her studies she created new textile materials such as smart textiles, textiles for lighting design and paper textiles. Additionally, she focused on the design theoretical investigation of textiles in architecture. After having had some touch points with the fashion industry, interior design and lighting design, she is currently working as a Colour & Trim designer in the automotive sector.

Contact
scholl.theresa@gmail.com
www.theresascholl.de

Charlotte Sjödell

'anyone can have an opinion and an idea but making solid proposals that can be visualised, tried and evaluated requires years of training'

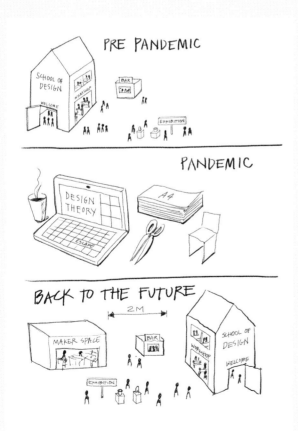

A

Who is the designer of the 21st century and what do they need to know? In difference to other professional titles, *designer* is not just owned by the educated designers but claimed by the many. Design, design thinking, and innovation is on everyone's lips these days. Is a five-year education really needed or is an online course in design thinking enough to practice design?

When I graduated as a designer from Art Center College of Design in 1999 I claimed to be either a product designer or an industrial designer. Industrial designer, I thought, well described some of the specific knowledge and design skill required to work with mass production. Many educators have moved away from 'industrial design' to just 'design'. 'Industrial' has almost become a dirty word and there seems to be a desire of removing oneself from anything related to *mass*. Mass production and mass consumption leaves a bitter taste and we long for small scale. This is in many ways a natural response to the changes we see in society.

The design occupation has broadened, and designers are now engaging in solving some of our times most pressing issues related to topics such as climate change, health, digitalisation, democracy etc. *Product* has more than ever come to include not only physical objects but systems and processes. Over the years we have seen an increased interest in understanding how the products we produce will impact society and the environment, both short and long term. Educators are struggling to introduce or extend topics to an already packed curriculum. We ask ourselves what is most important, and which are the essential skills of a designer. If we add something, what do we take away. Offering physical workshops for modelmaking and teachings is costly, and it is tempting to scrap the workshop, move to digital and up the numbers of students.

During the pandemic we are now given an unplanned opportunity to review what happens when we remove some of the physical aspects of design education. The students that have been enrolled in the program since several years understand the importance of physical models and have responded by making prototypes in their parents' kitchens and student dorms. Some first-year design students have never been in the building at their design school and have no experience of how to operate machines or have spontaneous discussions around the handling of an object, the tactile aspects of a material or its texture. What the effects of this will be, we might not know until some years from now.

It has become increasingly difficult for individuals to understand how products are being put together and how they are being made. Production is out of sight in another country and products are not made to be taken apart and be repaired. Kids spend less time around physical objects and more time as consumers attached to a glossy screen. During the pandemic many of us have realised the negative consequences of our screen time and have taken up new interest in the crafts or music to kill time in isolation. One of the few positive aspects of the limitations in our new lifestyle has been the time given to learn a new skill. We have become increasingly uncomfortable to take on anything that requires effort and focus over time. Time on task is one of the most essential aspects of learning especially in relation to skills that require some type of muscle memory.

Honing your skills in sketching and model making requires both think and doing and is nothing you learn over night. It is the combination of doing and thinking that makes design stand out from other fields. Thinking is simply not enough. Anyone can have an opinion and an idea but making solid proposals that can be visualised, tried and evaluated requires years of training. A lose concept or idea will change as you define it. When you visualise something, you are forced to define it and to take a stand. Knowing what good design is not the same as making good design. When moving into the future we must ensure that the doing is not lost and that the designer can offer a unique combination of skills and traits. When responding to change it is important that we do not lose our core. One visit to the workshop is simply not enough and an online course in design thinking will certainly not make you a designer, so let us hope we can soon get back to the building and do our thing.

Charlotte Sjödell
Charlotte Sjödell received her degree in Industrial Design at Art Center College of Design in Pasadena, USA. In 1999 Charlotte was hired by Ford Motor Company in the UK to work as a designer. When she left Ford in 2003 Charlotte was the Chief Designer of the European Colour and Materials Department located in Germany. She has over the years worked as a freelance designer working with projects for IKEA and other producers. Since 2004 Charlotte has worked as Senior lecturer in Industrial Design at Lund University, Sweden.

Contact
charlotte.sjodell@design.lth.se
https://www.lunduniversity.lu.se

A
Sketch by Charlotte Sjödell:
'Back to the future in design', 2021.

Olaf Redlin

From the seed to the 'Red Heinz' – worth to take 'time'

When the sun is still low during the day in the Riddagshausen monastery nursery in February and the rights are around freezing, Pauline can look forward to the delivery of the seeds for the coming spring. Pauline, a certain member of the Mehrwerk gGmbH (WfbM), is responsible for sowing tomatoes with her instructor. Very special free yourself on the 'Red Heinz'! An old seed-proof, also reproducible variety. Wholeness and with a lot of feeling, Pauline places the seed in the prepared peat-free BIO substrate bed and completely covers the seed with soil.

And now it's time to wait! Well, not quite. Pauline has to make sure that the seeds are still closed and can germinate. Then you can quickly marvel at the success of your work and a dainty green plant will see the light of day. After pricking, i.e. transplanting in a single pot at a pleasant 15 degrees in the greenhouse, the tomato plant stretches into the light and matures into a strong young plant by May.

Pauline is now selling 'her' plant, proud of her work, and by July at the latest it will have its first delicious rolls, which will then be harvested for salads, pasta or simply for snacking.

Contact
Olaf Redlin
Mehrwerk gGmbH
Klostergärtnerei Riddagshausen

Björn Stichnothe

Design Process in Disabled Workshop

Handicapped workshops are often used as cheap labor for the production of industrial goods, production is rarely cost-covering. The potential of the participants is often much greater than expected.

The idea is to help disabled persons to transform their expression into products. Disabled people have different perceptions in life and different levels of disability. An autist or a blind person have different selection criteria with regard to haptics. Haptics, curves and edges made by their hand are here more important, simplicity of their creations far away of Computer Aided Design (CAD).

Design as a part of integration and communication between each other or to change their point of view. Structures can be perceived and implemented differently.

Björn Stichnothe
Dipl.- Des. (FH)
HAWK Hildesheim Product Design Studies
Lighting consultant, mobile architect,
Production Manager @ Mehrwerk gGmbH
Product Designer

Contact
stichnothe@steg20.com

Jennifer Wetzel

Future design is experienced past, openness and curiosity

Didon's words 'higher, faster, further' lost their glory. Three words, that describe exactly all that we've been after for a long time. The dilemma? We lose focus and get lost in higher, faster and further. 'We' is all of us. And who will fall by the wayside in the hype of 'higher, faster, further'? Also, us. And with us, our competence to capture, understand and create.

Today, mankind has experienced both, analog and digital. In the post-digital era we live in, it is our job to combine and to use the experiences from the past. And thereby driving progress further. The children of the 90's have the luxury of having been born into a world that was still analog, albeit already transforming, and then growing up in a digital world. In the design of the future, our aim must not be to digitally replicate reality as precisely as possible. Rather, we should create a basis upon which we can act naturally and work digitally. The art is as old as humanity itself. Whoever wants to create, must know about the material. The carpenter must feel the wood. If you want to master glass, you must accept shards and other failures. Because failure with the material also drives the creator to the better, changes his movements – sometimes also the direction – and ultimately leads to perfection. We have to feel, smell, see and taste material to be able to grasp the object in its entirety. A chair cannot be built by someone who has never sat on one. For future design, this means, that we cannot create digitally, if we have no analog knowledge to begin with.

Part of my Master's degree in Clothing and Textile Technology at the Hochschule Niederrhein – University of Applied Sciences in Mönchengladbach, was a crocheting project with children in elementary schools. The idea of this project was to show that textile craftsmanship is extremely relevant for children of that age. Working with your own two hands is an indispensable part of development. To learn, to experience, and to draw creativity from it. The pride and admiration you get from a tangible object in your hands, created by yourself, is independent of age and culture. Most certainly, it is of tremendous value for you, the creator.

Speaking of future design scientists, it is my wish that we understand that aesthetic education begins with infancy. That our children learn to use their hands and their minds for analog and digital development. And that we recognize that both will become their tools one day, and we should help them train those tools equally. Ultimately, when we talk about design engineering of the future, one thing cannot be missed: the human being with his or her hands and real-life material to create things.

Jennifer Wetzel
Founder and CEO of Wachsling GmbH
Jennifer Wetzel is product developer and textile engineer. She lives together with her husband and their son with another family on an old farm in Dormagen. 2018 she graduated at Hochschule Niederrhein in Mönchengladbach with the thesis 'A beeswax-coated natural fiber fabric for packaging food – founding a start-up as an approach to reducing aluminum and plastic waste within the 2018 environmental debate'. After completing her studies in 2018, she founded the sustainable company Wachsling GmbH, today one of the leading German beeswax wrap producers.

Contact
hi@wachsling.de
https://wachsling.de

A

Wenche Lyche

'playfulness in creative design processes will be 'the new black''

B

Manual exploration versus digital exploration is nowadays part of students learning processes and there is a growing need in society to understand the relationships between manually created surfaces and creativity based on new technology. Still, I claim that manual exploration is necessary to understand and cherish the unexpected in exploring what may come that can be understood, transformed, appreciated and create new value, for product design students and the product itself. Outside winter workshops, gave students the opportunity to use forces of nature in their search for new surfaces design and then digitally transformed into patterns for digital print on textiles.

Provoking textiles into new shapes by using minus 10-15 degrees gives students the possibility to be free and present while working and experience "joy of not knowing" and "the joy of uncertainty as a gift" like Joanna Macy calls it, as opposed to more result driven processes with digital tools. Physical active workshops also create arenas for communication, critical thinking and discussing new insights in design in variable locations such as outside workshops in cold winter days. Questioning when the silk organza is no longer soft but stiff and cold? Sunlight shining through the silk giving contrasts and intriguing aesthetics. What about the printed wool standing up straight and the surface in frozen condition is suddenly shiny?

Playfulness in creative design processes will be "the new black" and create opportunities for students for excellent and unique design results as an opponent to the ever commercial marked with mass production. This means, it is of great importance that creation for uniqueness in manual exploration in product design education gives a better sustainable awareness in surface textile design for interior and fashion.

Wenche Lyche
Associate Professor
My background is from tailoring of exclusive womens-wear for private customers running my own brand. My main focus as a designer, is natural fibers, design-lines challenging body language and customization. In my career I have participated in several national and international exhibitions with my objects of jackets made only of zippers and these are in the permanent collection at the Nationalmuseum in Oslo and Trondheim. I tutor Bachelor and Master students in product and fashion design at Oslo Metropolitan University and hold workshops in exploratory surface pattern design at Estonian Academy of Art. My responsibility at Oslomet is the textile workshop where I tutor students in the world of textiles and our digital textile printer with manual exploratory methodology.

Contact
wely@oslomet.no
Instagram: https://www.instagram.com/lycheli2/
Oslomet: https://www.oslomet.no/en/about/employee/wely/

A
Frozen wool on frozen lake.

B
Organza with frozen preformance.
Pictures by Wenche Lyche, 2021

Giulia D'Aleo

'toward a future of sustainable interactions'

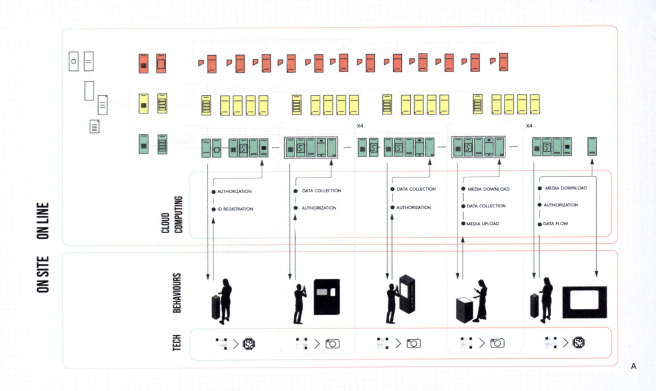

A

The way human and machine think is not alike and there are still many gaps that prevent them from understanding each other. Nonetheless, if we want to pursue the Digital Transformation this friction must be overcome with a double action: we need to focus on the design of humanity-centered machines empowering us, as much as on the expansion of our consciousness about the way they can affect our lives. The design field offers several practices geared toward dealing with complexity that have proved to be useful to the definition of desirable future scenarios for human-tech interactions. For instance, System Oriented Design offers a pragmatic approach to transform the daunting complexity that comes with it, into an opportunity for inherent innovation. It is used to dissect and analyze all of the parties involved and, by fixing their relationships, finally improve the overall system. Thanks to this process we can also improve the sustainability of interactions by supervising each passage form an ethical, economical, social and cultural point of view. On the other hand, Speculative Design, with it's critical approach, can be used to arise questions about a problem instead of solving it tout-court. It creates narratives of future realities to assist the exploration of unwanted side effects not yet visible in the present, therefore allowing for pivots and redirections of the future. It has already shown to be useful to address the implication of the diffused presence of technology in human life, rather than its mere application. At the Master program in Digital Interaction Design I'm attending at Politecnico di Milano, I'm discovering how we can manage to take this considerations into the design of sensible interactive products, responsive environments, multimedia interfaces and personal artefacts. We are asked to improve them by putting into practice some of the aforementioned principles, and I reckon that our projects always benefit from the debate with experts in fields different from design. I'm looking forward working in more and more heterogeneous teams to break boundaries and find unexpected connections that can result in a more fluent dialogue between humans and machines.

A
Diagram of the system oriented approach used for the definition of a mobile experience aiming to help user get a better understanding of the contents of an exhibition. Graphic by Giulia D'Aleo, 2021.

Giulia D'Aleo
Master student of study program Digital and Interaction Design at Polytecnico di Milano
After a classical education pursued in her teens, which formed her way to look at the world complexity, she moved to the design field. At Politecnico di Milano and during a semester at Berlin IU, she studied interior design and learnt how to build environment that enables interactions among human and objects. At the moment she is studying Digital and Interaction Design to be able to expand her ability to design and study smart objects and the way they can sustain human growth without menacing it.

Contact
giulia.daleo@mail.polimi.it

Valentin Wachs

'how to attack the root'

We all agree, that humanity is facing a lot and will be facing even more problems in the future. Some of them smaller, some of them bigger. But the problems in detail don't have to be discussed to find somewhat of a solution. Because to solve a problem, you have to find the root of it. So where do all the problems come from? The clear answer to that is: Human behaviour.

Every problem humanity has is self-inflicted. So how do we change human behaviour? Again, simply by finding the root. This is where it gets interesting. Because the two key factors regarding this are upbringing and surrounding. Changing the behaviour scheme of the parents, which will be mirrored by the kids, would be difficult. But there is a point that change could start from. Which brings me to the main message I want to elucidate: The education system, which is a big part of our lives, is the important factor we need to renew.

If we would teach the new generations how to reflect, act and bring change, the world would be a much better place in the future. We should for example be thought in practical economics, sustainability, problem-solving and mental control, just to mention a few topics that need to be discussed. Those elements could be the base that we build on.

Valentin Wachs
19 years old, Germany

Sibylle Klose

In recent years, we have encountered a remarkable shift towards a much more skill- and competence-based education that shapes experiences and pattern of differences. Interconnecting different contexts and practices, interweaving various approaches, enable us to create broader, fluid as flexible bodies of experimental knowledge. This is considered as initial, crucial groundwork to generate creative, innovative responses to today's issues and intended futures. Alongside, it requires the most empowering spaces allowing vulnerability, doubt and failure to foster and cultivate each other's mutual learnings. Since we do not own ideas, they must be dis-covered, un-covered or even re-covered anew by a prepared, elastic mind embracing dualities, uncertainties and the unknown.

Sibylle Klose
Sibylle Klose, Professor for Fashion, Dean for international Affairs, Course Director BA Fashion at Pforzheim University School of Design. Based in Paris, she worked as a fashion designer and design consultant in France, Italy and Germany including luxury brands like Thierry Mugler and various Fashion Houses of LVMH. She honed her international expertise in teaching, research and program development by serving various institutions including École Nationale Supérieure des Arts Décoratifs, the Institut Français de la Mode, former Parsons Paris School of Art and Design and Parsons Paris the New School in Paris, Shanghai University – International Institute of Fashion and Art in Shanghai, and as consultant at Parsons the New School in New York. She has launched several new bachelor and master-level programs.

Contact
https://designpf.hs-pforzheim.de

'shaping experiences and patterns of differences – to dis-cover future ideas'

Maximilian Krummen

'making the world
	a tiny step better with my art
and performing, creating:
	awareness, will and trust'

My core of making music / performace

'Music can bring peace /
brings souls together'

> Making music / singing /
> experiencing a live performance fills
> the body with endorphins
> Making music together strengthens
> the awareness of being part in a
> bigger group

'Before you sing, hear what you sing'

> 'Playing the inner game is a matter of
> developing three skills: awareness, will
> and trust. These skills in turn help us
> to achieve relaxed concentration, the
> 'master skill' that allows us to balance...'
> (from Barry Green with Timothy Gallwey
> 'The inner game of music')

'Be in it with all of your heart'

> „He never denied his character: to cling
> with all his soul to what was in front of
> him: hoc agebat. He didn't rest until he
> realized what he wanted to learn.
> He did so with languages, arithmetic,
> mathematics, algebra...'
> (Constanze Mozart about her husband
> Wolfgang Amadeus)
>
> Be prepared to listen to you soul, before
> you say something or create something,
> no matter what.

'Ah, tutti contenti saremo così' /
'Ah, we will all be happy like this.'

> *From the finale of
> Mozart's Le nozze di Figaro*
>
> Music helps us to forgive and find
> respect for each other
> The combination of music / lyrics /
> performance can lead us to our inner
> feelings, can access our soul.

Maximilian Krummen
Classical (opera-)singer Baritone
I am a classical singer in opera and also for art-songs and oratorios. As a vocal artist I not only try to entertain my audience, but also to encourage all of the above in them. I hope to make the world a tiny step better with my art and performing. I truly believe that music and text as well as the associated performances or visual arts can affect other people to hope for a better and sustainable future.
Rostock, September 13th 2021

Contact
www.maximiliankrummen.de

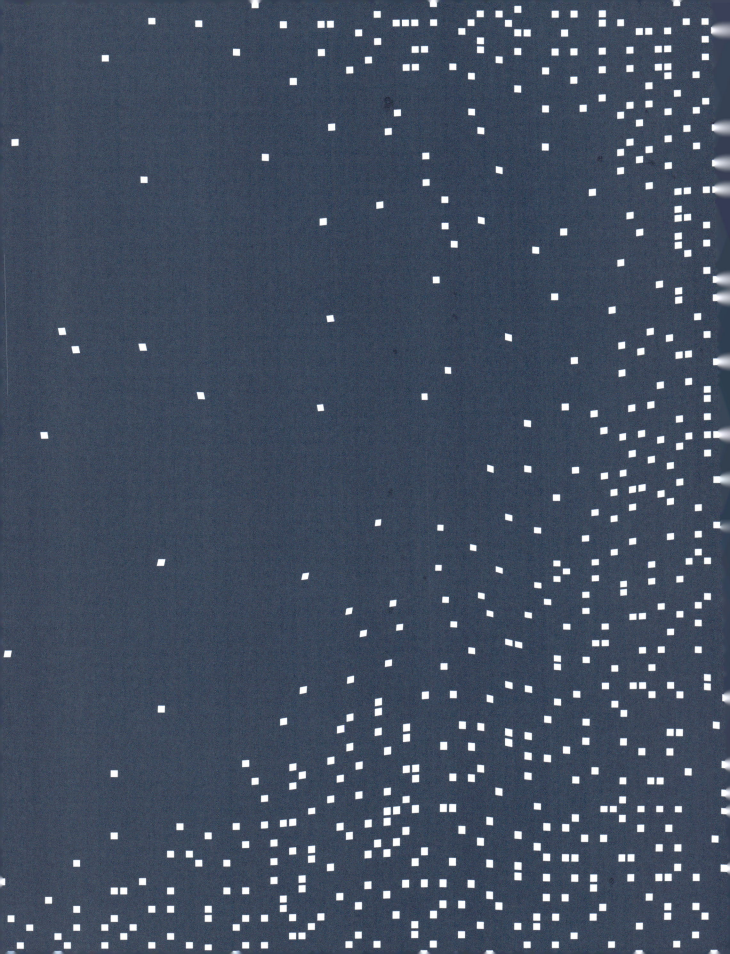

CHAPTER 07

246 — 267

'Sketching' sustainable future together in a hands-on, non-hierarchical and holistic design engineering education culture – cherishing playgrounds for art and design engineering heritage by using 'Knowledge Banks'

7.1 A position paper for interacting and integrating systems in sustainable design engineering and design education – state-of-the-art, holistic, non-hierarchical and interdisciplinary learning

Keywords
Collaborative engineering, to know the history of knowledge to build the future, evolution of (media) behaviour, memorability – creation – learning – ideating, motivation to act, osmotic learning landscapes: all stakeholder together, design trainers under the radar, mentoring as curriculum part, The big three 'E's in design engineering education, inharmonious management – disruption as wishful thinking, sketching to incorporate knowledge.

This postdoctoral research study opens the mind to more 'European togetherness' in design engineering and interlinked, non-hierarchical interdisciplinary activities. It showcases the benefits of a holistic approach to cultural education – from interdisciplinary scientific and practical projects for pre-schoolers all the way to retired experts. This statement is interlinked with an innovative designing format for the future. The consequences related to the shift in design, which stem from a shift in cultural behaviour and designing processes, are causing systems- and thinking-oriented design methods to be used. At the same time, it documents the evaluated results of the representative case studies and best practise models. These studies are **collaborative engineering and product design education** projects that took place within the initial European cultural spaces – Germany, Great Britain and Scandinavia – during the 2010s.

In the same way that the shift in design – in other words the 'design turn' – is related to other 'cultural turns'[94] (Bachmann-Medick), we may also consider the ongoing paradigm shift that is happening as part of the fourth (digital) industrial revolution. We cannot ignore the effects it is having on working and designing conditions in industry and in educational models, as part of this interconnected evolution. As such, we have to discuss how living conditions and interactions with objects and things, or with thoughts and thinking models, relate to materiality and immateriality, and increasingly abstract phenomena.

[94] See Bachmann-Medick, D., 2016, Cultural Turns, De Gruyter; ibid: Schatzki, Theodore R. et al, 2001, The Practice Turn in Contemporary Theory, Routledge.

In design theory, the use of arguments related to sociological, philosophical and cultural anthropological circles – which provides a broader perspective of design doing and design theory – is based on working with reference sciences (Bezugswissenschaften). The precondition of the methodological comprehensive strategy in design theory relies on arguments derived from the empirical and hermeneutical use of tools applied in field studies, as well as process related thinking models, SOD (Meadows), and epistemic-based analyses. This makes it possible to achieve design solutions and design concepts through co-designing that are increasingly theoretical – in other words immaterial.

As we begin the 2020s, the Internet of Things (IoT) and the Process of Things (PoT) are both mirroring the use of knowledge in a globalised, interlinked culture – a culture of knowledge. But a critical analysis of the developing process, forces us to look back in time, to understand the usage, as well as cultural and material behaviour, in relation to immateriality and behaviourism today.

Design history is usually interpreted in terms of its ethnological and anthropological significance: That means simultaneously understanding, interpreting, and evaluating design engineering education history, in order to ideate! Design history was used as element of design theory, to evaluate the usage of **design objects**, which was less based on questioning design engineering '**education** history', or '**design processes development history**'.

The history of knowledge provides answers in design and design engineering. This helps to understand the **evolution of behaviour** – as well as the uses of design objects and concepts throughout time, as they relate to social, cultural, and of course national political developments.[95]

According to design theorist Claudia Mareis, historically evolved knowledge cultures and concepts are based on a discontinuity of knowledge: they use historically important breaks, voids, and disruptions to build up knowledge and transfer it into new solutions[96] (see Mareis, C., 2011). In 2020, this comparative perspective once again in using social fractures[97] (see Wachs, M.-E., 2008), underlines the existing design process and the rethinking of systems, in order to transfer knowledge into new design solutions. Thereby, the aim is to find 'good' usages of design objects and concepts – ones that are sustainable, long-lasting, high-value and accepted. That means, we have to look at the design model and its historical development. The conventional designing model (process) is analysing objects or processes, comparing these with other systems and demand, thinking in systems, developing ideas from different perspectives (various experts and interested individuals), to ideate visions and translate them into real-world solutions, evaluating from different point of views, evaluating people's demands and synchronizing them with industrial and economic conditions, and ultimately defining future design engineering artefacts that can be presented, communicated and discussed again.

With the ongoing development of designing and education models, the method of **co-designing** and the reduction into four core components, by co-creation of 'analysing – understanding – ideating – evaluating'[98] (see Sanders, Elisabeth B.-N., et.al., 2008) came into focus, starting

[95] See van den Boom, Holger, 2010, Das Designprinzip. Warum wir in der Ära des Designs leben, Kassel University Press; and others by Holger van den Boom and Felicidad Romero-Tejedor.

[96] See Mareis, Claudia, 2011, Design als Wissenskultur – Interferenzen zwischen Design- und Wissensdiskursen seit 1960, transkript, p. 379.

[97] See Wachs, Marina-Elena, 2008, Material Mind – Neue Materialien in Design, Kunst und Architektur, Dr. Kovač, p. 219.

[98] See Sanders, Elisabeth B.-N., et al., 2008, Co-creation and New Landscapes of Design, research gate: DOI:10.1080/15710880701875068

around 2008. This method provides excellent opportunities to integrate different stakeholders' interests for defining sustainable solutions in a great variation of disciplines.

Currently, we can observe an evolution of the term *designing*. It could now be defined as the joining together of creators' skills and the benefits offered by taking a holistic view. This can be achieved through non-hierarchical, diverse teams of experts, talents, pupils, teachers, civil and institutional representatives, who all participate in the creative process and learn from each other to serve the needs of the future. This is the reason why the term 'design', and its significance needs to be reviewed.

The impact of historical perspectives on the term 'design', is relevant to the change in usage in former times and today, as it relates to the current 'design shift' and the present change of object use and cultural behaviour: for example, from a renaissance perspective of art and art science, or by looking at Leonardo Da Vinci's practical creations and Georgi Vasari, as a theorist; or the use of sketches to represent, develop and discuss design; or by taking into account that we are living in complexity based on Simon and Norman and supported by other design theorists, like Felicidad Romero-Tejedor, Holger van den Boom, Claudia Mareis and others contemporaries; and lastly the act of **'rethinking of systems and technology'** (see Braham W. W., et al, 2007) – with the aim of **managing the process and being open-minded towards the unexpected!** Designing by means of reflection, and using design theory as tool, always begins with the interplay of an object, a concept and a system – just as it was in the past, as Norman underlined in 2011: *'In the end, the rules all evolve around communication and feedback. The design must include appropriate structures to aid human* **comprehension and memory as well as tools for learning**, *and for handling unexpected events [...].*
The system might be used alongside other systems that do similar things but follow different design principles, and even though each alone might be sensible and understandable, [...] **the design has to support usage in the face of the unavoidable interruptions of life'**. [99]

Now, more than ever, this statement relates to the discipline of sociology from the designer's view point, in the same way that Richard Sennett referred to the 'flexible man' (Sennett, R., 1998) within the system of capitalism and the 'corrosion of character' in 1998. The changes in working and living conditions (see chapter 3 and 5) continue to shape human life on earth, including in during this epistemic and holistic era. As a result, we are facing challenges related to the shift in our progressive construction of the future cultural mindset within the scope of the fourth industrial revolution taking place today, which can be resolved by means of constructing (complex) systems in real-time with the help of technological possibilities – integrating an 'inhomogenous management' (Wachs, M.-E., 2022) in accepting: friction, failure (see chapter 4) and disruption as whishful thinking to 'develop' the system. At the same time, Norman's statement supports the idea of linking creative methods to the psychological effect of entering into a state of 'flow' (see chapter 6.1), in which unexpected rather than pre-constructed system-based learning takes place in reflective and creative environments. This involves collecting information, experiences, and

---

[99] See Norman, Donald, 2011, Living with complexity, Birkhäuser, p. 224.

background knowledge, which can be transferred spontaneously to new situations and can also interrupt parameters within the design process by simply taking a break.

Creativity needs breaks. This means using the skill of **memorability in the moment of *forgetting*** elements of the cultural memory (see Esposito, E., 2002). Generating innovative concepts is possible in moments of relaxation and thus forgetting. Ergo, applying proven structures to new fields requires a moment of forgetting. Abstract thinking is enabled by means of interlinked neuro-physical systems in our brain, with the help of past experiences and that merge with knowledge in certain unique circumstances[100] – that is important for every human being.

We can evaluate memorized aspects within the framework of our culture and its cultural memory. In addition, we can reflect on design history and how design objects were used in the past, as well as how the design process in education and industries evolved. This is all connected to our knowledge of culture and cultural experiences – including our experiences with 'otherness'. It represents a valued ethnological interplay of global cultural aspects that question ethically and morally correct behaviour and attitudes towards otherness, which is not only reflected through design artefacts or their usage. At the same time, **otherness means welcoming foreignness**, and can inspire design and influence the outcome of design processes. A global product language in design reflects the zeitgeist and a modern use of design: Design means using and learning from other disciplines – for example, by using the method of field studies based on the discipline of ethnology. Applying design theory means reflecting, evaluating and constructing artefacts, while mirroring the cultural memory of the past and present. This helps to formulate scientific methods in design and questioning the discipline to come: design science – in ongoing vivid discussions (see Wachs, M.-E., 2010).[101]

As previously shown in chapter 6.1, 'welcoming otherness' is important when using Integrated Cultural Intelligence (ICI). This model focusses on the skills essential for managing the changes we face in the 21st century. With regard to the 'theory of cultural intelligence', by Earley and Ang, *'Three general components capture theses* (note: in managing intelligence) *skills and capabilities: cognitive, motivational, and behavioural. Without all three of these facts acting in concert, a person does not display cultural intelligence.'*[102] In our current knowledge-based society, neuroscientists and psychologists are intensively observing how cognitive skills are trained, and sociologists and cultural anthropologists are examining skills in behaviourism. The trigger points studied in philosophy, like demonstrated, are more focused on the **motivation to act**.

If we look more closely at the latter, we learn that the chance or potential failure to motivate a member of a knowledgeable society within a complex world system is tied to the moment of action. Corine Pelluchon is questioning, how can a philosophy of corporality, which extends an ethics of vulnerability whose key category is passivity, be reconciled with the social contract, of which many thinkers, have said that is was a snare and a continuation of domination by other means? The philosopher is arguing, that this contradiction evaporates once we understand that there

---

[100] See Wachs, M.-E., 2008, Material Mind – Neue Materialien in Design, Kunst und Architektur, Dr. Kovač, p. 210 and 243 f; ibid: Esposito, E., 2002, Soziales Vergessen, Suhrkamp; ibid: Stern, Elsbeth and others.

[101] See Wachs, M.-E., 2010, in Romero Tejedor, F. and Jonas, W. (ed.), 2010, Positionen zur Designwissenschaft, Kassel University Press, p. 201.

[102] See Earley, Christopher P. and Ang, Soon, 2003, Cultural Intelligence. Individual Interactions Across Cultures, Stanford University Press, p. 59 and p. 67.

is no opposition between sensing and intellect, the heart and the reason, force and fragility.[103] (see Pelluchon, C., 2019 (2015))

In other words, **even if we wish to change, will we? If so, we have to face the cultural values of our complex system, and the following key (timeless) questions:**

What are the values of our time? How are they defined and followed? With regard to the culture of remembrance, where do these values come from? How significant is the cultural impact of a common or culturally determined existing set of values, or a more individually based set of values? And, like Donella Meadows asked in relation to 'thinking in systems': What causes a person or a society to give up on attaining 'real values' and settle for cheap substitutes? How can you key a feedback loop to qualities you can't measure, rather than to quantities you can?[104]

Things brings us back to cultural values and intelligence. As Wilhelm Krull and other authors have underlined, it is not a question of 'quantity' in terms of the right measure for excellent cultural education (and systems) in revalue 'quality' in academic circles.[105]

Instead, the quality needed to measure excellent education, could be generated through the use of 'integrated cultural intelligence' (see chapter 6). Also, a rethinking of the system (see chapter 2, 3.1, 5.1 and 6) is necessary to relay values and technical possibilities within today's fourth industrial revolution. If the current chances and challenges related to change rely on an optimal combination of contradictions to be combined in the best possible way – like: AI + EQ, design thinking + action painting – we also have to rethink technology alongside sociological values as they progress over time.

We must also look at the systemic framework of cultural education. The term 'culture' inherently includes living conditions, and how people live with objects and industrial processes today. These are being modified by AI-influenced production processes, digitalized CAD models that cooperate with learning machines and robots, which are combined with human skills. This scenario describes state-of-the-art processes that are currently a work-in-progress in the design engineering industry. The powerful influence of artificial and robotic systems is facilitating a paradigm shift for the global economy and society, as well as for the public and individuals. Like Braham and Hale reference design in architecture, **a model of 'rethinking technology'** is needed for all applied art disciplines: *'[…] architecture and technology or, more precisely, the tools of design and construction, have become a matter of systems. […] Ultimately, rethinking technology and architecture in the age of systems means rethinking the practical and ethical dimensions of change, development, and evolution in architecture.'*[106]

The paradigm shift we are currently experiencing is expressed through trailblazers who implement the use of 'unequal systems': e. g. combining logical and emotional skills, and implementing analytical systems and digital tools that are supported by high-tech AI solutions for optimizing living conditions in a complex world. Felicidad Romero-Tejedor refers to acting as 'thinking designer' and being something between a scientist and an artist. By this it means to design with an emotion-based, more 'intuitive and synesthetic style of thinking'. So, a designer is someone

[103] See Pelluchon, Corine, 2019, Ethik der Wertschätzung, wgb, p. 352.

[104] See Meadows, D., Wright, D. (Ed.), 2009, Thinking in Systems, p. 169.

[105] See Krull, Wilhelm, 2017, Die vermessene Universität, Passagen.

[106] Braham, William W. and Hale, Jonathan A., (Ed.), 2007, Rethinking Technology. A Reader in Architectural Theory, p. xiii.

who creates through thinking and the creative act – in design doing. Within the culture of sciences in the 21st century, designers are able to deal with unknown and undefined things, and they are able to step into new foreign fields while working in interlinked, non-linear thinking processes. According to Romero-Tejedor, the designer possesses interlinked, non-linear thinking, that is necessary for creating accommodative systems.[107]

As such, by interacting with and integrating systems into the design engineering future, it is paving the way for meeting new needs in various industries and in engineering – needs related to the fourth industrial and social revolution, **and** thus also education. Developing concepts for innovative interlinked design theory within complex non-linear economic and common social systems requires design thinking. Digitally based tools are used in analogue linked systems, and are influenced by emotionally triggered human behaviour. At the time, this creative (ideating) system **is in the process of evaluating and revaluing the sketch as a tool for representing and visualising ideas by materialising concepts.** (see 'immersive designing' in chapter 7.2.2.)

As illustrated in chapter 4.2 – according to the method of material-based design thinking, as well as other methods like Ewenstein and Whyte[108] are formulating, – the **sketch** (also in architecture) could be considered an epistemic-based scientific object used as an interlinked, collaborative creative tool for visualising ideas. The use and value of sketching and drawing, as a unique and original tool in design, is currently being re-assessed. In addition, the design code, or product language, in design is being transferred and communicated via media. This always relates back to the cultural imprint and sociological determination, on the one hand, and the anthropological significance of the culture of remembrance (Erinnerungskultur), on the other. *'Materials, which determine the terms relating to design objects, also supplement the materiality of culture. [...] This specific material code is always a part of social behaviourism and can thus be decoded within that cultural context.'*, like Wachs underlined in 2008.[109]

Hence, taking socially and politically relevant action in design, combined with the rethinking of technology and systems, in order to achieve sustainable design engineering processes and education models as part of the paradigm shift(s), **requires a focus on European forces and fragility, and demands taking a strong position towards sustainable design solutions.**

---

[107] See Romero-Tjedor, F., 2007, Der Denkende Designer – Von der Ästhetik zur Kognition. Ein Paradigmenwechsel. Olms, p. 188, translated by Wachs.

[108] See Ewenstein and Whyte in: Mareis, C., 2010, p. 70 f.

[109] See Wachs, M.-E., 2008, Material Mind / Materialgedächtnis, Neue Materialien in Design, Kunst und Architektur, Dr. Kovaç, p. 255, translated by Wachs, 2021.

Accordingly, this post-doctoral thesis, presents to various partners (stakeholders), **a position paper about generating sustainable design solutions – for economies, education and life conditions** – which can be achieved by:

- enabling design engineering to take place in cross-cultural, interconnected digital and analogue working spaces
- designing in cross-generational courses that incorporate all ages – from retired experts (vocational education) to children
- engineering the future by promoting artistic education
- encouraging children in pre-school to take on responsibility
- creating non-hierarchical learning landscapes (osmotic spaces)
- offering pre-school design activities – please take this seriously – and creating integrated design(ing) playgrounds
- understanding and acknowledging the impact of art and culture in everyday solutions
- including sustainability and design engineering in school curricula (in Germany!)
- promoting cleverness in interdisciplinary connected thinking and creating – but you need to provide a knowledge base
- using AI (Artificial Intelligence) and EQ (Emotional Intelligence) to create a new attitude, and rethink moral correct values for 2030
- incorporating 'Integrated Cultural Identity' (ICI) in a new form of co-designing that respects 'otherness'
- using current events as socially relevant design tasks ('event of the day as a trend')
- materialising immateriality by visualizing through materials and the new model of iDIM (interlinked Design Information Modelling)
- applying design theory as method of 'sketching the world in the next Augmented Reality (AR)', through analogue or digital hands-on design
- sketching to incorporate knowledge
- managing your 'Knowledge Banks' in the context of knowledge management.

7.2 Rethinking design systems and codes

7.2.1 The paradigm shift in design and 'media behaviourism' – and its consequences

Beside the 'must haves' previously outlined in this position paper, in terms of sustainable engineering design (SED) for the future, this post doc thesis gives insight into the changes in the process of managing design objects AND concepts. The change in cultural behaviour and indeed of 'media behaviour' (Wachs, M.-E., 2020)[110] has been important for the ongoing paradigm shift in design(ing), as it relates to the development of technology and natural objects. **Even though the 'sketch' is being used to incorporate knowledge, it is also serving as a communicative representation tool – for example, a materialised sketch as a 3-D mood board that is filled with meaning. Sketches are used to incorporate knowledge of design** and have a positive effect on outcome, because sketches originate in open-mindedness, and visionary, discursive questioning. It does not provide final definition, but utilizes a more intuitive evaluation and measurement system that is expressed by the interaction of style elements: line, colour, outline, surfaces, traces, ductus, as they relate to the individual person – the designer – and the 'design didactic approach', that can be stated. (see graphic 05, p. 157)

Thanks to this change in behaviour, a concomitant acceptance of design in the 'art business landscape' is obvious: In the past, designer objects were regarded as key historical artefacts, a collector's passion for classical furniture, for example. Always considering form and function, as well as high quality, aesthetic and long-lasting design language. The highest level of which was demonstrated, by inducting highly valued members into the halls of fame of museum collections (e. g., MAY DAY lamp by Konstantin Grciç at MOMA).

Today, design is part of the paradigm shift and is accepted as an integral part of society in Europe – not only as a highly valuable expression of culture, but also as an element of social responsibility and a strategic management tool that influences everyone's habits.

[110] Note: media behaviour is a new term by this study, related to the anthropological point of view of 'material behaviour', defined by Michal B. Schiffer and used by Hans Peter Hahn during the 1990s; ibid: see evaluated at the first book by Marina-Elena Wachs, in 2008: Material Mind.

The paradigm change is demonstrated by two great lines: We must take responsibility (more than just economically): Firstly, we need to incorporate private capital for all entrepreneurial undertakings. The new term 'Skin in the game' – as used in 2019 by different authors (see e.g. Taleb, N., 2008) – is a highly valued and increasingly important necessity on a meta level. This phrase is mentioned by Taleb and used by Affentanger, and refers to taking responsibility for your actions – your invention. Secondly, tapping into a more playful character, with designing landscapes, for example, makes it easier to create the right atmosphere for designing an ethically acceptable future – imagine to combine the method of design thinking + hands on experiments within the 'flow'.

'as 'failure is a natural and desired part of the innovation process' (see Ashley Hall), 'disruption' is to evaluate the process, and for inharmonious management a wishful thinking'

(Wachs, 2022)

7.2.2 Attaining interlinked Design Information Modelling (iDIM) through mixed media design doing AND physically apply tactility, textiles and light in engineering processes

Perhaps now is precisely the right moment in time for a new sense of corporate social responsibility to emerge and be enabled by designing together in immersive, integrated AR systems, which are combined with an interlinked Design Information Modelling (iDIM) (see Wachs, M.-E. 2020) system, developed in the near future.

It is possible to draw benefits from different design systems: for example, architecture is moving forward in applied art disciplines. This was shown in the recent past with the use of Life Cycle Assessments (LCA) and Building Information Modelling (BIM)[111] and the help of 'integral planning'[112]. That should also be at the forefront of design engineering in 2022, although it is rarely used in enterprises or at universities in Europe at the present time.

Aside from a willingness to create open-minded designs in collaborative, immersive multi-reality spaces, the future of design engineering, requires technical interoperability and a common technical understanding and usage. After all, the final products depend on how good the design engineering experts' competencies are, as they are the people who create the digital tools and software designs. They also manage the digital libraries that allow people from around the world to connect with other experts, as well as with these tools (see chapter 6.2 in detail). Ergo, **the design system – and thus the sustainable design engineering solutions – can only be as good as the experts' skills are**. Accordingly, from a **technical perspective** we need:

- an optimised software to link different workflows, working methods and forms of presentation
- to interconnect and manage the workflow of mixed cross-cultural design engineering teams, while respecting the need for empathy and understanding towards different ethical *product* languages
- dynamic labs (and further developed 'tablets') with understandable logic and intuitive controlling and acting

- software designed by German or European enterprises, to guarantee the US or Chinese market will not take away rights and research sources
- to construct tools in accordance with the discipline: in design engineering a more sculptural, 'all in/skin in the game' construction in AR
- to combine interactive design information management and interlinked collaborating design tools.

---

[111] See the statement by Architect Moritz Schwarz, p. 214 ff.

[112] See Oechsner, A. E. (ed.), 2020, p. 22 ff; ibid: 10 ff et al.

Note
See Wachs, M.-E. synaesthetic workshops during the 1990s until students exercises in Master programme today; e.g. – for children at buntich school for art: painting 'through' music; or for Hochschule Niederrhein at each winter semester programme: designing / sketching through music. one example: project: design is music is design – European united, 2018: https://www.hs-niederrhein.de/faculties/textile-and-clothing-technology/research/designmusicdesign-eng/

From the **human perspective** and the belief that human capital is our best resource for ideation, we need:

— a greater acceptance in design engineering in Germany, to reap the enormous benefits that cultural education offers when disciplines and senses are combined: synesthetic designing underlines the multifunctional design and resilient use (circular economy)
— to embrace and take advantage of collaborative design projects, for example, with a lighting planner and a musician working together in a virtual room (a form of AR), who can use VR glasses and controls to influence light, tone and tactile (and textile) impulses in the future

— greater self-confidence in following and trusting our own intuition in design engineering
— to trust in the senses and insights of full sensory experiences (German: sinnliche Erlebnisträger) – a synesthetic education
— to take steps in improving cultural education: FIRST, analogue hands-on design, with manual *grasping* (be/greifen, er/fassen) of artefacts and natural objects; SECOND, digitally sketching the world.

Mixed Media design doing in AR + to be 'touched' by physical interaction, is an idea that was initiated by Marina Wachs, to be discussed and driven by a mixed team of experts, in August 2020. The team comprised industrial designers, lighting planners, students and movie and software experts in Hamburg[113]. (End of 2021 re-discussed by Wachs and team of Roland Greule.)

The idea was to create a more sculptural and tactile design tool, influenced by parameters of light, textile and tactile value. This was discussed in July 2020 with various stakeholders, who all offered different perspectives: from the students' view, to the alumni, and teachers, as well as the lighting planners, fashion brand entrepreneurs and other fashion creators, and textile design engineering experts. The discussion was made possible via a virtual video conference tool, a consequence of the covid-19 pandemic in 2020, we worked in digital rooms, thanks to Ulrike Brandi, Markus Rindle, Urs K. Rouette and some talents of the mentee group.

Afterwards, an interdisciplinary discussion, organised according to the designers' point of view (Wachs), was further fuelled by sharing best-practise models of taking action and designing in the virtual lab with IT specialists and textile engineers.

Aiming for the incorporation of more tactile, immersive, virtual sketching and drawing, we organised an interactive design experience at the VRHQ lab in Hamburg, Germany, which took place in August 2020.[114] Here, experts from different disciplines came together, and we followed the trend by looking at a variety of VR glasses and analysed what the next product generation needs: e. g., we began with glasses made by Oculus (quest model) because of the low price, but discussed the users' acceptance and potential preference for Holo Lens 2 glasses, made by Microsoft.

[113]
VRHQ lab Hamburg, 27th of August 2020, Wachs, M.-E., Ulrike Brandi GmbH HH mit Malina Angermeier, Susanne Ahmadseresht, VRtualX lab Hamburg, Roland Greule, FTZ Lab HAW Hamburg, Department Medientechnik Professor für Lichttechnik, Simon Graff, nextreality Hamburg, Mathias Wolk, CEO bei VRtual X GmbH und realTV group GmbH & Co. KG.

[114]
Thank to Prof. Roland Greule, director of FTZ institue of HAW Hamburg, link: https://www.haw-hamburg.de/ftzdigitalreality/

QR Code 09

[115] Movie QR Code 09 'Textile Engineering in the future', see EEE conference 2022, Wachs, Balbig, Grobheiser, Scholl.

At that moment, the cost of Microsoft's model was six times higher than that of Oculus. A more advanced interlinked system used with VR glasses has been tested in the automotive industry, some by Tesla as well as in particular the system made by Seymourpowell and the HTC Vive Pro Glasses that use special controllers.

We were able to get more 'skin in the game' with the help of VR glasses and controllers, but were still looking for more physical sketching and editing possibilities in finding a balance for analogue combined digital (sketching) and manufacturing in textile industries.[115] (see QR Code 09)

From the perspective of textile and industrial designers, this type of designing in a more physically interactive and interdisciplinary environment could be beneficial when triggered by textiles, this would be a more tactile-driven process that could be combined with lighting and smart software with future interlinked design information management tools. This involves providing all design engineering partners with the tools and design documents they need for CAD sustainable solutions, while respecting all media rights (text, picture, sound, etc.) – a process that goes beyond tangible tactility (and in memorizing different cultures).

'The state-of-the-art virtual designing-management today is, sketching' the vision behind it, integrating more tactile elements and corporeal experiences.' (Wachs, 2021)

At this moment in time in the realm of DESIGN, Virtual Reality VR and AR are mostly used for presenting, communicating, informing, managing data, and marketing, that is demonstrated in graphic 09[116], but not as a (physically interactive) constructive design engineering tool that can help ideate new objects and concepts for a sustainable future.

The following project process demonstrates (and should be) the state-of-the-art method, used for building a hierarchy of competence (levels) in the digitalised world of business, in the year 2022:

Graphic 09
State of the art virtual designing management, Marina-E. Wachs, 2021.

[116] Wachs, M.-E. and Ahmadseresht, Susanne, product developer at VRtualX lab Hamburg, 27th of August 2020.
https://nextreality.hamburg

5. Interactive playing and designing – sketching the future in virtual spaces…

4. Project collaboration through till training the trainer

3. Library completion (functions / colours / surfaces…)

2. Exploration sketch and virtual configuration

1. Product presentation

To conclude, the three most primary tasks that need to be executed for the future in design, with **interlinked** virtual and analogue spaces in Europe (and around the world), are:

— to design more sculptural, perhaps more free-form based, construction in AR?
The pre-requisite for this is to train the best educated people in hands-on design – sketching and drawing in an analogue manner… that means 'revaluing' art in educational programmes in pre-school

— finding respectful ethical solutions for using data and determining the rights of German /European software
— to develop digital skills **by questioning how and why to use** artificial AND emotional intelligence (AI and EQ).

If we revisit the headline **'rethinking design systems and codes', we can conclude, that human beings are the starting point and the gauge for using mixed media in cross-cultural spaces when designing. This means including more (virtual) physical interaction and trust in the ability of people to use the controls intuitively, while still respecting a good and organised iDIM (interlinked Design Information Modelling): one with room for the 'flow', for breaks, for voids… for failure!**

The paradigm **shift in media behaviour taking place now, in the 2020s, is restoring the worth of human manual skills and is leading to an evaluation of the possibilities offered by digital tools and AI 'in real time'.** The use of AI must include the acceptance of an ethical best-practise approach to the artificial system. Consequently, this means respecting and seriously investing in, and integrating emotional intelligence into the process, for the benefit of European education and economic systems.

'to precise for the textile business: We are entering the future of balancing 'fashion tech' and hands on fashion design, that will result in a 'smart haute couture' of the 21st century, that will result in new educational programmes'

(M.-E. Wachs, 2021,
note: 'fashion tech' refered
to Troeters, Marina, 2020)

7.3 E-Valuating European design memory by using a 'Knowledge Bank' – research responsibility in Ethics, E-technology, Education (the big 3 E's for cultural education and economy)

Keywords
Design as a personal or collective memory; the practitioner as a historian; the historian as a practitioner; design history as memorable design archives, mapping design history, design memory and the Knowledge Bank in Design.

[117] Müller-Funk, Wolfgang 2010, Kulturtheorie, A. Francke, p. 291, term translated by Wachs, 2021.

[118] Note: this could likewise be compared and underlined by the neuroscientific perspective by Gerhard Roth, Gerald Hüther and by cultural scientist Wolfgang Müller-Funk; see Roth, G., 2003, Aus Sicht des Gehirns, p. 89 ff; ibid: Müller-Funk, W., 2010, p. 299;

[119] Note: since the 1980s we have been using the term 'product language' in design and associated applied art disciplines. If we generally look at the historical, anthropological and cultural studies, one of them has to be in focus: In the 1980s the 'Cultural Memory', as defined by Jan Assmann and the work of Aleida Assmann prove how important the relationship is between 'text, tradition, and the way of thinking', in relation to the culture of writing. See Assmann, Jan, (1992), Das kulturelle Gedächtnis, Schrift, Erinnerung und politische Identität in frühen Hochkulturen, C.H. Beck, p. 280. ibid: see Assmann, Aleida, 2018, Der europäische Traum – Vier Lehren aus der Geschichte, C.H. Beck.

Knowledge is the greatest resource in Europe, and the only source the Germans have for creating economic and ecological profit. Knowledge is fundamental for obtaining the skills you need to become an expert in various design fields and for interdisciplinary learning as part of a holistic educational background – a pre-requisite in the 21st century. Without these complex design landscapes to provide continuous learning possibilities, young designers will not be able to compete and be a part of the European and global design community in the future. Using knowledge for ideating innovative concepts and objects means **remembering, recollecting and combining your archived information in the most effective possible way: Remembering** can be expressed by means of speaking, and discussing with the help of **3-D based sketching (in AR) – by materialising ideas**. Material things are used to represent, like a 'mis-en-scène', or the staging of props and elements. In conclusion, remembering means visualising, discussing, narrating and evaluating in a special form: **Thoughts are materialised – a phenomena where materials 'speak'.** It is a form of expression **that uses the language of media, and narration is considered from the perspective of linguistics**, yet based on cultural theory and subsequently design theory.

Thus, we must also consider what this narration includes. Beginning with the philosophical point of view, Aristotle's view of narration has been interpreted by the cultural theorist Müller-Funk as a *'normative handling instruction'*.[117] Nowadays, in 2021, narration is represented by the following triad: *experience – storytelling – remembrance*[118], which ultimately includes usage (instructions) and 'design doing'. In general, the way of using design objects and interpreting design languages relates to texts, objects, traditions, rituals and the way of thinking in a specific cultural circle.[119] Therefore, the people expected to use the objects will have to connect with the past and former cultural elements before they begin using digitalised media and AI-based aids in the future, in order to truly understand these new systems.

However, a certain amount of care must be taken in this process. Just as memory shapes behaviour and can be used to help people understand systems, it can also be used to control them. 'Memory is the weapon', the title of the story published by Don Mattera in the 2007, is an example of how weaving together different disciplines can have an impact. He states:

'it weaves together both his personal experience and political development.' (Mattera, D. in: Ruhkamp, U., 2019). During the 1950s, in South Africa, this was also the title of 'Robin Rhode's art exhibition, 'Memory is the weapon', held at the Kunstmuseum Wolfsburg, Germany, in 2019. Ergo, if we regard memory as weapon, it is clear how this claim can be interpreted on a political level: It illustrates how the story associated with a period in time and its culture can be compared to other developments, and can be used as a political weapon, via art or artificially created objects, like design.

Equally, if textile designs or **woven constructions in design theory focus on media and materialising visual concepts, we can consider material as a weapon. In the same manner, media** is a weapon today in all businesses, as well as everybody's private life. Hence, our materials and communication media have to be humane and sustainability for the planet's sake.

Thus, it is a question of identification **and** identity, and using artificial intelligence **and** emotional intelligence in a common environment to help people remember values. It is worth designing valuable objects, concepts and systems for the future – things worth remembering in 100 years that will serve as part of the 'Knowledge Bank' (Wachs, M.-E., 2020). By using existing knowledge archives, we are inevitably constructing the next layer of 'cultural memory' (Assmann, J., 1990s), the next 'material mind' (Wachs, M.-E., 2008), or the next 'media mind' (Wachs, M.-E., 2021).

Is it important and does it matter, whether we use artificial intelligence (AI) or emotional intelligence (EQ) for the purpose of generating the next 'minds'? Does it matter, whether we use digitalised, immaterial ideas or 3-D materialised ideas that can build a tangible future? Yes, it does, because the capability to analyse, understand, ideate and evaluate sustainable solutions will be essential in the circular economy, and subsequently in the post-digital era within a decentralised economy.

It should be noted that these capabilities will depend on the ability of humans to connect and exchange ideas in the real world. In the end, **you must evaluate past results to determine the next steps towards creating an ethically valuable education in design engineering**. As we have learned, such evaluation is based on different languages and 'handling instructions'. Handling instructions are an essential tool in didactic of art and will be in didactic of design in the future. In turn, these are based on tradition, text and images; on the mediation of contemporary moralistic values through rituals; on a fundamental Knowledge Bank that is hidden in objects, concepts, stories and archives (personal archives and the archives of society). The function of an idea that has been materialised is *'parole'*, according to Ferdinand de Saussure, which refers to (scientific and public) discourse as part of rule and ritual.[120] **Language is coded and trained. Language is the power of culture, like it is in product language. Thus, in design theory it is fundamental to analysing, understanding, ideating, evaluating, communicating and creating cultural and media behaviour.**

In 2020, we celebrated the 100th anniversary of the memorable 'Golden Twenties' in Europe. 100 years ago, new design gave expression to a new life – a desire that was particularly strong after the 1st World War. The new shapes of objects and materials supported the new style, found in fashion, furniture and other forms of design: three-dimensional objects for

---

[120] See Foucault, M., 2012 (1971), Die Ordnung des Diskurs, (l'ordre de discourse), Fischer.

everyday use. The focus on form in industrial design mirrored the demands and economic benefits brought about by industrial processes that aimed to optimise the availability of cultural goods and increase prosperity in European societies, during the Golden Twenties (the German equivalent to the roaring twenties in the USA).

If we compare this to the ongoing 4th (digital) industrial revolution and the impact of artificial intelligence, the concurrent celebration of technological advances seems obvious. We can use the 'Knowledge Bank' as a memory archive, by transferring this knowledge into new materials and forms that are being produced by fab labs and 3-D printing processes, for example. Although this new era of technological possibilities also unveils new demands associated with the Internet of Things, we simultaneously have to respect the social revolutions that are calling for sustainable behaviour and adopt circularity in design concepts and production processes: One 'slow design' solution could be to provide users with unique products manufactured using high-quality processes. Producing a limited number of unique objects for users, offers a way to slow down consumption for a better world – reminiscent of former lifestyles.

So, before society shouts 'memory full', we must re-evaluate our cultural behaviour with regard to materials, human resources and the way we produce and transport things around the world. This research thesis highlights examples of high-value manufacturing processes, combined with smart design solutions conceived in analogue as well as digitally interlinked working design landscapes. This creative process requires us to 'remember and forget': First, we must use the memorable craftsman skills found in the Knowledge Bank and merge it with the smart, high-end, state-of-the-art technology that is available now. Second, we must follow the neuropsychologists' advice and accept that 'creativity needs breaks', i. e. allowing some breathing space for solutions to be forged spontaneously. These solutions form part of the memorable act within the design process. This goes back to the 'flow' in design, as described in chapter 6.1, meaning 'wishful thinking' – letting thoughts and wishes flow – which demands space and time in which to forget what 'is', in order to discover the new ideas that 'can be'.

This describes the creative power that a moment of relaxation holds, in which the absence of memory allows room for remembering essential elements of design – and not only material(s). **Looking back at design history – as well as the history of technology, and the history of design engineering education** (see chapter 2 and 4) **– it gives us insights that allow us to enhance, to interrogate and evaluate designing systems for the future. It is a way to express design history as a memorable design archive.** This archive helps to shape morally accepted behaviour. Both entrepreneurs and researchers, who are responsible for generating ethical value in life conditions, can use this to create a socially responsible economy.

In consequence, EQ, as it relates to AI systems, must be used in design Engineering processes, by integrating human skills for generating Ethical working and living processes, expressing the big 3 E's in education in the future:
Emotion – Engineering – Ethics. (see graphic 10, p. 267)

The vision is to find new ways of working together in different fields – diverse cultural education levels, non-hierarchical working structures, collaborations between a diverse range of talents (teachers, experts and entrepreneurs and children) – while respecting children's diverse needs and aspirations, as well as the concepts developed for researching 'demands'. When questioning how 'designing conditions' will be in the future, we must consider the pure technological pre-requisites, on the one hand. But, on the other, we must question how to train future generations, to ensure that we are giving them the ability to solve complex problems by using engineering tools in an ethically correct way: respecting 'otherness' of culture and cultural behaviour. – This is also welcoming people with handicap, and allow them for integrating into the design process. – It is a question of providing quality, holistic cultural education – which includes knowing the history. Using design in an unconventional yet historically defined manner means the practitioner is also acting as a (design) historian and the historian as a practitioner. This refers to design doing from the very first scribble or sketch onwards, and using this as a basis for analytical discussions and epistemic considerations. It is also used to transfer knowledge of the process to the product and system, which always relates back to the cultural background and imprints contained in an interlinked design engineering community. In terms of education systems, it is obvious that, in order to draw greater benefits, we must develop more holistic design education methods that are based on interdisciplinary design engineering – ideally in 2022!

In order to achieve this, **we must (re)build the 'house of design engineering', due to the paradigm shift** caused by the 4th industrial revolution, by creating innovative structures and ideating construction systems, with the help of *'téchne' (Ancient Greek: παλιά)* and *'theoria' (Ancient Greek: θεωρία)*. This means, learning and internalising ethical values – with regard to cultural, social and political issues – and expressing these outwardly.

Let us look of the metaphorical image, that Vilém Flusser meant in 1993, that although this type of home construction represents a technical revolution that reaches far beyond the skills found in architecture and design. This kind of architecture – without roofs and walls – open to the world... would change the way we live. People would have nowhere to hide, they would have no ground to stand on, or walls to lean on. They would have no choice, but to hold one another's hands.[121]

In conclusion, we can say that ethical, sustainable design doing is in the hands of Europeans, waiting for them to ideate and evaluate design engineering solutions together within well interlinked learning landscapes, as outlined in this study.

121
See Flusser, Vilém, 1993, p. 81, translated by Louise Huber-Fennell, 2021 'Allerdings wäre so ein Häuserbau eine technische Revolution, die weit über die Kompetenz der Architektur und des Design reichen würde. (Das ist übrigens der Fall bei allen technischen Revolutionen.) Eine derart dach- und mauerlose Architektur, die weltoffen stünde... würde das Dasein verändern. Die Leute könnten sich nirgends mehr ducken, sie hätten weder Boden noch Rückhalt. Es bliebe ihnen nichts übrig, als einander die Hände zu reichen.' (Flusser, Vilém, 1993, p. 81 and reprint in Entry paradise: 2006, p. 159).

7.4 Design Engineering – sustainable and holistic: education as 'industry' – a place for education and to remember!

Design trainers 'under the radar' and design theory as a method are influencing design engineering and education

As we enter the 2020s, working with interdisciplinary design theory means epistemic-focussed designing: in other words, analysing and showing true understanding and respect for 'otherness' – or foreignness. This design approach makes it possible to ideate through the use of an epistemic-based synthetical process that is viewed as a valuable resource and optimises understanding of how to gestalt a reasonable life on earth – described as 'Weltentwurf', in German.

Within this frame, focussing on the term 'industrial design engineering' as it was defined, in the years after World War II and at the beginning of the paradigm shift based on 'good form', we can see that design and production processes were based on *interlinked* production and design culture, that evolved in Europe. Not only the accessibility of production processes was based on physical infrastructure and the exchange of products, machines and know how. Until recently, the accessibility of design culture was also influenced by the physical presence of teacher, trainer and students – particularly for the Bauhaus, for example – and a bilateral exchange between European countries. After the second World War, cultural understanding and empathy for the cultural habits of the other foreign nations played an integral role in this. Thus, cultural accessibility was made possible through applied design history related disciplines. It follows that design history and design objects of the past are part of *Europe's memorable culture*. And, although the design objects of the past may have gone unseen, they are not absent.[122] **We have to remember and discuss the design history of objects, concepts and education, in order to evaluate and synthesise future design concepts. Design history tells us important stories about economic, ecological, social, cultural and political aspects – despite their physical absence.** Being a relevant aspect of design theory, design history delivers essential epistemic input in every present discourse – **not only through a continuum of design forms but also by allowing us to recognise disruption and voids in history, opening our mind to new cultural experiences**. The discourse also relies on individuals at design schools,

---

[122] See Hugo, Victor, 1865 (1985), Actes et paroles. Pendant l'exil, in: Assmann, Aleida, 2018, Der europäische Traum – Vier Lehren aus der Geschichte, C.H. Beck, p. 190.; ibid: Novotny, Helga, 2005, Unersättliche Neugier – Innovation in einer fragilen Zukunft, Kadmos. Engl: Insatiable Curiosity – Innovation in a Fragile Future.

design theorists and teachers to set the course. Some influencers include Gui Bonsiepe and Tomás Maldonado, Herbert Lindinger, Holger van den Boom, Bernhard Bürdek, Egon Chemaitis, and – from the more recent past – Felizidad Romero-Tejedor, Brigitte Wolf, Kora Kimpel, Claudia Mareis, Sabine Foraita, Wolfgang Jonas, Marina-Elena Wachs, Ashley Hall, among others (... and new talents to come like Theresa Scholl, Giulia D'Aleo).

Wilhelm Braun-Feldweg, is perhaps one of the lesser-known design teachers and coaches for good industrial design. He discussed the term industrial design, at that time a very new and unconventional term, as it related to history and ongoing design processes. Braun-Feldweg also demonstrated the significance of revolutionary industrial development, in the context of social development and the dependency on the technical possibilities available in 1966. He had a very open-minded perspective: *'Everything that is generated is a product that results from the perceptions, opinions and conditions of diverse provenance.'*[123] In the 1960s his view was new to industrial design. Accordingly, you may consider that everything produced today is a result of the perceptions, opinions and conditions of diverse heritages. Nevertheless, 'Provenance' is currently undergoing a revival within art discussions in different cultural circles and will also be your task in design in the future.

Our cultural memory is generated from memorable skills, knowledge and technical possibilities, as well as the human ability to connect and synthesise the questions regarding the future. But this all takes place within the context of, and connection to, the past. As we move forward, this involves reflecting upon and rethinking material and immaterial worlds and behaviourism. Designing the next cultural memory and material mind means understanding, respecting and expressing empathy towards otherness. It means welcoming foreignness, which takes us back to the beginning of this research thesis and the initial thesis and thoughts, which are supported by the philosophers Paul Ricœur and Corine Pelluchon.

In consequence, the change required for a valuable sustainable future must be designed by tapping into the **motivation to take action – design doing in a corporeal manner**. The concept of using design theory as an active designing method means incorporating the need for a discourse about Europe's commemorative culture, as demonstrated through the heritage shared within the EU, since 2014. As we have learned, designing the future means remembrance, while simultaneously allowing enough room for space and breaks – or voids – in which the memory of absent design objects and concepts can be recalled and used as a valuable Knowledge Bank for OERs (open education resource). 'Forgetting' is part of the creative process. **We must use remembrance – the 'lieux de mémoire' – in the design process to discuss and revisit the past, so we can create the future.** This also means taking responsibility in design engineering, whether for globally applied solutions or education. The European (design) identity is based on 'regulations and moral principles' created by the four lessons taught by history[124]: to protect freedom and democracy, and to respect human rights and the culture of remembrance. (see Assmann, A.).

---

[123] Braun-Feldweg, W., 1966, Indusrial Design heute – Rowohlts deutsche Enzyklopädie, p. 185: 'Alles, was entsteht, ist das Produkt aus Anschauungen und Bedingungen verschiedenster Herkunft.' translated by M.-E. Wachs, 2021. By the way, today Benita Braun-Feldweg cares about Wilhelm Braun-Feldweg's heritage, not only by the formation of the only German design award of design theoretical textes, which is calling every 3 years. see: http://www.bf-preis.de/start/index.php?page=preis

[124] See Assmann, Aleida, 2018, Der Europäische Traum – Vier Lehren aus der Geschichte, C.H. Beck, p. 185 ff.

'Erinnern beruht auf Erzählung (Narration), auf Erfahrung, Erkenntnis, Epistemologie'. = 'Remembering is based on telling stories (narration), experience, cognizance and epistemology', as Marina-Elena Wachs formulated in 2020 and 2021, for different events, talks and workshops (see submission for Design History Conference 'Memory full', (Basel, CH); see talk at Free University Bozen (Bozen, I), within the topic about 'Sketching – significance in art, design and management of sciences'). We have to evaluate the design historical narrations, in order to define the criteria for good design in the future, as well as for good designing conditions and education. For sketching the future of Design Engineering's history, 'yesterday's future' begins right now and it is worth the investment in holistic pre-school education, that suits the shift in 'industrial' performance today.

It is about your own (!) attitude towards design engineering and culture, in terms of sustainable design engineering education and 'industries' – even though the term 'industry' is in the process of changing right now. However, this is another story, for another day which we do have be aware of. And, last but not least, it is about your own (!) attitude as design trainer heroes – perhaps as a catalyst in mentoring the 'next' mentor – as design engineers and trail-blazing entrepreneurs: the future of designing with AR is in your hand including the pre-school playground with your own ductus and an holistic view.

Graphic 10
Emotional Intelligence –
Engineering / Education –
Ethical Value, E-Core Elements,
M.-E. Wachs, 2021

QR Code 07
See further development and information via conference presentations and talks the last years via movie by M.-E. Wachs

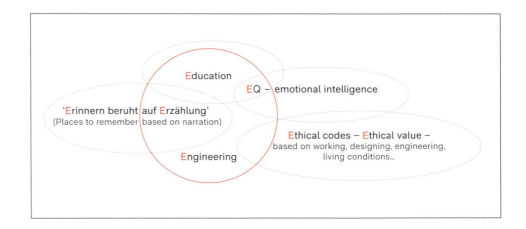

Epilogue

Once a person begins to study Industrial Design or Design Engineering, their academic interest focusses on receiving a contemporary, competitive training as a designer and engineer, and in fact, to complete their portfolio with a final project that contains about 20 % innovation – design with the future in mind.

A new students' fresh outlook is not based on the heritage or the history of industrial design education – let alone dedicated to their entitlement to a didactical and history-based training. Their interest in design history generally increases over the course of their studies, once they are introduced to their first design and art history classes. This often helps them develop an awareness of past form/ornament language, product language or a set of materials, which become part of their pool of knowledge that can be drawn from during their studies and for years to come.

A wish for future education programmes is to recognize the concept of a 'Knowledge Bank' (Wachs, M.-E.), which comprises general 'collections', such as form collections and university archives, as well as immaterial archives. This pool of information represents a chance to provide a 'cross educational' perspective, and also serves as an essential element of the Europe Project. This collection and knowledge resource consists of both material and immaterial elements, representing cultural memory. It promotes learning about form collections, aesthetic abilities, form sensibility, proportionality, language ability and understanding, as well as the ability to present arguments and process abstract ideas. These skills – along with design reflection, process knowledge and intercultural competencies – help to better understand the system of designing, which will shape future design attitudes. These material and immaterial archives support the evolution of design shapes and artefacts of our future design history. They also spark – even demand – debates and discussions relating to both design theory as well as hands-on design, and will lead to new social (analogue and digital) design solutions. Hopefully this will help an in-depth understanding of tomorrow's 'Design Science' evolve.

In the future we can create a design engineering education model that is characterised by the following elements:

— design engineers will be complex problem solvers
— sketching the future will focus on hands-on designing in Augmented Reality and…
— …'material based sketches' by 'materialising immateriality', this begins at pre-school, which include synaesthetic play full designing and art experiences
— …sketches will showcase the significance of drawing in the future, by increasing self-confidence and self-consciousness …
— …this means: the essential skill will be 'thinking through sketching', and this will overshadow 'drawing for production'.

On the one hand, this model means 'thinking by sketching', which emerges by 'materialising immateriality' (Wachs, M.-E. 2020). On the other hand, this model promotes the reciprocation between cultural education systems, industry and society – sustainable and holistic.

Picture 09
Marina-Elena is sketching within the Augmented Reality, 2021, in VRHQ lab Hamburg, Germany, thank to Roland Greule!
https://www.haw-hamburg.de/ftzdigitalreality/
picture M.-E. Wachs 11/2021.

Picture 10
HANDS on design, Sketching the future together, knowing about the right moment in time to take hands on or digital tools.
Art by Marina-Elena Wachs sketch: of 'we have to look more precise series' since 2018, here: 'digital nerd 1', 2019, Venezia.

'you have to regard the past, to DESIGN a sustainable future – towards to more European togetherness in design engineering'

Marina-Elena Wachs

Pic. 09

Pic. 10

Acknowledgement

I would like to thank so much, to all my international cooperation partners, friends and family, students and alumni – the next design engineering partner, for discussions, reflections, in calling critical educating competences, challenging my capabilities and for growing together in design engineering and scientific business fields – in different expert fields the last decades. Because precondition is essential, to build up creative competences and the ability for reflection with an open mind set, that is shaped as flexible but at the same time with recovering knowledge foundation all the time. It gives us the possibility to answer with bright design skills, methods and applications to the planet's demands!

First of all, I have to express my graceful thank to my mentors, for the unconventional ('unprätentiös') support along the last decades and valuable 'time' of inspirations, discussions and talks together: Ulrike Brandi and Michael Schwarz.

My special thank – named in alphabetical structure – goes to: Armand, Werner Aisslinger, Knut Amtenbrink, Claudia Bei der Wieden, Karin Behrens, Ellen Bendt, Hannes Böhringer, Holger van den Boom, Ulrike Brandi, Benita Braun-Feldweg, Claudia Bruns, Egon Chemaitis, William Franke, Ashley Hall, Klaus Hardt, Susanne Hauser, Wolfgang Jonas, Kora Kimpel, Sibylle Klose, Erich Kruse, Christian Labonte, Scott Lipinski, André Franco Luis, Karin Müller, Markus Rindle, Helmut Schlotterer, Theresa Scholl, Michael Schwarz, Joachim Schelper, Sonia Sin and Frans Schrofer, Gunnar Spellmeyer, Ulrike Spengler, Hans-Joachim Walitschek, Dorothee Weinlich, Gabi Wicke, Manuel Windmann.

I would like to thank the organizers, thinkers, motivators of EPDE conferences all the years: it was and is a great motivation and wonderful inspiration for me, and very worthwhile (learning) space for reflecting the present and the past with all design engineering partners for our ongoing – responsibly managed – future.

With graceful thank to all my local, European and international cooperation partners (some over the last decades), represented by the statements by invited experts around all creative disciplines, across cultures and across generations for creating our future together, to a turn for a better.

Thank you so much, grazie mille, sinceras gracias, vielen Dank, hjertelig tak, uppriktigt tack, tusen takk, merci mille fois;).

Special thank to MW and VW and to Isabell-Carola!

Best Marina-Elena

Imprint

© 2022 Marina-Elena Wachs, Braunschweig
www.marinawachs.de
© 2022 for the texts with individual
authors, photos see picture credits
© 2022 av edition GmbH, Stuttgart

This work is subject to copyrights.
All rights are reserved, whether the whole
or part of the materials is concerned, and
specifically but not exclusively the rights of
translation, reprinting, reuse of illustrations,
recitations, broadcasting, reproduction
on microfilms or in other ways, and storage
in data banks or any other media. For use
of any copyrights owner must be obtained.

Bibliographic Information published
by the German National Library.
The German National Library lists this
publication in the German National
Bibliography; detailed bibliographical data
are available on the internet at http://dnb.de

TEXTS AND DESIGN/GRAPHICS
Marina-Elena Wachs
ENGLISH LECTORATE
Louise Huber-Fennell
CONCEPT AND DESIGN
Eva Finkbeiner
TYPEFACE
RM Neue by CoType Foundry
PAPER
Magno Natural 150 Gr
PRINTING
Schleunungdruck, Marktheidenfeld
PUBLISHING AND DISTRIBUTION
av edition GmbH
Publishers for Architecture and Design

With special thank to Hochschule
Niederrhein – University of Applied
Sciences and the 'Gleichstellung
Kommission', for the idealistic support
of my international engagement.
We have done our best, about
the delcaration of data rights.
If anything different, please contact
Marina-Elena Wachs, thank you.

avedition

av edition GmbH
Publishers for Architecture and Design
Senefelderstr. 109
70176 Stuttgart
Tel. 0711 / 220 22 79 0
Fax: 0711 / 220 22 79 15
www.avedition.com
sales@avedition.de

Printed in Germany
ISBN 978-3-89986-362-8